PLEASURE, PROFIT, PROSELYTISM
British Culture and Sport at Home and Abroad
1700–1914

PLEASURE, PROFIT, PROSELYTISM

British Culture and Sport
at Home and Abroad
1700–1914

Edited by
J.A. MANGAN

FRANK CASS

First published in 1988 in Great Britain by
FRANK CASS AND COMPANY LIMITED
Gainsborough House, 11 Gainsborough Road,
London E11 1RS, England

and in the United States of America by
FRANK CASS AND COMPANY LIMITED
c/o Biblio Distribution Center,
81 Adams Drive, P.O. Box 327, Totowa, NJ 07511

British Library Cataloguing in Publication Data

Pleasure, profit, proselytism: British
  culture and sport at home and abroad,
  1700 – 1914.
  1. Sports – Social aspects – Commonwealth
  of Nations – History
  I. Mangan, J.A.
  306'.483'09171241   GV706 . 5

Library of Congress Cataloging-in-Publication Data

Pleasure, profit, proselytism.
  Bibliography: p.
  Includes index.
  1. Sports – Social aspects – Great Britain – History.
  2. Sports – Social aspects – Great Britain – History –
  19th century.   3. Sports – Social aspects – Great Britain
  – Colonies – History.   I. Mangan, J.A.
  GV605.P58 1987   796'.0941   87–24977

ISBN 0–7146–3289–9
ISBN 0–7146–4050–6 pbk

Typeset by Ann Buchan (Typesetters), Surrey
Printed and bound in Great Britain by
A. Wheaton & Co. Ltd, Exeter

CONTENTS

NOTES ON CONTRIBUTORS

Derek Birley is the Vice-Chancellor of the University of Ulster. His publications include *The Willow Wand*, a study of the mythology of cricket.

David Brown is an adjunct professor with the Department of Physical Activities at the University of Regina, Saskatchewan, following a year as the University of Queensland Post-doctoral Fellow in Brisbane. His research interests focus on the rationalization of sport in Victorian and Edwardian Canada and Australia, particularly in their private schools.

Richard Cashman, Senior Lecturer in History at the University of New South Wales, has edited two volumes on sports history and published two books on Australian cricket crowds and one on Indian cricket. He recently completed *The Demon Spofforth: Australia's First Cricket Hero*.

John Daly is Principal Lecturer in the School of Physical Education at the South Australian College of Advanced Education. He lectures on the history and sociology of sport and has published three books and numerous articles. Since 1973 he has led the Australian athletic team, including acting as head coach of the national team at the Montreal, Moscow and Los Angeles Olympics.

Richard Holt is Lecturer in History at the University of Stirling. An expanded version of his Oxford doctoral thesis, entitled *Sport and Society in Modern France*, appeared in 1981. Since then he has been working on a general history of sport in Britain since 1800 (Oxford University Press, 1988). He is an editor of *The International Journal of the History of Sport*.

J. Thomas Jable, Professor of Physical Education at the William Paterson College of New Jersey, is currently researching the role of sport in the urbanization of nineteenth-century Philadelphia. His previous work has explored such topics as Sunday sport, high school athletics and American professional football. In 1987 he completed a two-year term as president of the North American Society for Sport History.

John Lowerson is Reader in History at the Centre for Continuing Education, University of Sussex. He is an editor of the *International Journal of the History of Sport*, and the author of a number of books and articles on the social and economic history of leisure and sport.

Kathleen Moore is a Staff Tutor in the Division of Physical and Health Education at the Queen's University of Belfast. She studied physical education and history at the University of Alberta, and recently completed her doctorate at the University of Queensland.

André Odendaal is attached to the History Department at the University of the Western Cape in Cape Town. He completed his PhD at Cambridge in 1984. He is the author of *Cricket in Isolation: The Politics of Race and Cricket in South Africa* (1977) and *Vukani Bantu! The Beginnings of Black Protest Politics in South Africa to 1912* (1984), and has played first-class cricket in both South Africa and England.

Mike Speak recently became Director of Physical Education at the University of Hong Kong, after fifteen years in the Department of Physical Education at Lancaster University. During this time he completed a Master's degree at Liverpool on the history of French physical education and sport, and is now working towards a PhD on the history of sport in the north-west of England, where he was born and brought up.

Brian Stoddart is Senior Lecturer in Sports Studies, Canberra College of Advanced Education. He is the author of *Saturday Afternoon Fever: Sport in the Australian Culture* (1986) and (with Ric Sissons) *Cricket and Empire: The 1932–33 Bodyline Tour of Australia* (1984), as well as numerous articles on sports history and culture in the British Empire, after fieldwork in New Zealand, Australia, India and the West Indies.

Wray Vamplew is Reader in Economic History at the Flinders University of South Australia. He has published widely on sports history including *The Turf* (1976) and *Pay up and Play the Game* (Cambridge University Press, forthcoming). He is editor of *Sporting Traditions* (the journal of the Australian Society for Sports History) and a member of the international editorial boards of the *Journal of Sport History* and the *International Journal of the History of Sport.*

Gareth Williams is Senior Lecturer in History at the University College of Wales, Aberystwyth. He has written widely on the social history of sport and popular culture in Wales in the nineteen and twentieth centuries, and is the author (with David Smith) of *Fields of Praise: The Official History of the Welsh Rugby Union* (1980).

ILLUSTRATIONS

ACKNOWLEDGEMENTS

The editor and publishers gratefully acknowledge permission by Marylebone Cricket Club to reproduce the photograph of 'The Winner of the Match' by Henry Garland. Acknowledgements for other illustrations are also due to the British Museum; John Daly; *Punch*; *Allan Glen's Monthly*; *Western Mail*.

Doris Mangan is thanked for her care and patience in compiling the index.

INTRODUCTION

Leisure, as Peter Bailey has suggested, is now seen belatedly in its own right 'as a significant element of social experience, whose history is of particular importance in the broader exercise of reconstructing the kind of lives lived by the ordinary people of the past' (*Leisure and Class in Victorian England*, 1978, p. 1). The same claim may now be made with equal force for sport, yet historians have taken an inordinately long time to appreciate its relevance for the lives of both the influential and the insignificant of past communities. And, as Richard Holt has pointed out, historical matters of far less importance have their serious recorders and commentators and, it might be added, their academic conferences and literature. Failure to get sport into sharp academic focus is, of course, a form of intellectual myopia born of long-established prejudice. Such short-sightedness is slowly being corrected. Eric Hobsbawm recently argued that sport was one of the most important new social practices of the Europe of the late nineteenth and early twentieth centuries, and as such played a central role in the creation of politically and social cohesive 'invented traditions' (Eric Hobsbawm and Terence Ranger (eds.), *The Invention of Tradition*, 1983, p. 298).

Pleasure, Profit, Proselytism is concerned essentially with sport as a cultural phenomenon in Britain and its imperial territories, lost and retained, from the period of the Industrial Revolution to the Great War, and Hobsbawm's contention is confidently extended in this volume beyond Europe to these possessions. The main purpose of this exploratory work is to set aspects of the sport of this imperial community within the culture in which much of it arose and within the cultures of dominion and colony to which much of it was transmitted. As Charles Tennyson once reminded us, the British taught the modern world to play, with varying degrees of success, in idioms of their own confident construction – social, moral and technical. The contributors in the following pages deal with a variety of consequential issues: most importantly with the role of sport in the establishment and maintenance of class-consciousness, and, as a corollary, in the creation and reinforcement of class differences at home and abroad; with sport not only as a source of tension, as one writer has put it, outstripping the reach of existing systems of social control, but also as an intended instrument of hegemonic influence utilized in self-assured attempts to achieve adherence to middle-class metropolitan values; and finally, and paradoxically, with sport as a means of establishing virtually simultaneously class, national and imperial unities.

At one and the same time, it is suggested, this book reveals that Robert W. Malcolmson's 'vacuum theory of the growth of leisure' (*Popular Recreations in English Society 1700–1850*, 1973) is an over-simplification, that subsequent to industrialization in Britain the common people demonstrated a clear capacity to 'make their own culture' and that a belief in a downward process of cultural osmosis in matters of physical recreation is both naive and inexact. It also offers support for Hobsbawm's views, in a far wider setting than European society, that sport in the late nineteenth century provided a means by which the bourgeoisie took up class positions against the masses, acted in part as a replacement for religion in the creation of complete cohesion and served in a larger context – once it was institutionalized – as a symbol of national self-assertion (*The Invention of Tradition*, pp. 298–303). While fresh material is to be found here against which earlier theories may be ranged, new theoretical perspectives are also set out for review.

Ground is broken empirically – Birley, Brown, Daly, Jable, Lowerson, Moore, Speak, Stoddart and Vamplew substantially, but not exclusively, place their emphasis here – and conceptually, as exemplified by Holt's concern with proletarian self-image through modern sport and with soccer as a form of urban 'festivity', Mangan's suggestion that sport in late Victorian Scottish middle-class education was a calculated attempt by the 'colonized' at national (British) embourgeoisement and regional (Scottish) deracination in the pursuit of acceptance and prosperity, and Cashman's assertion that cricket and colonization should be viewed 'from below' as well as 'from above' if the story of its role in imperial cultural diffusion is to be recounted adequately. And while various invented traditions associated with sport in Motherland and Empire are examined by these writers, traditional mythologies associated with sport are exposed in Williams' account of the disjunction between image and reality in the attempted creation of 'Welshness' through rugby football, and in Odendaal's investigation of the extent of black imperial middle-class athletic activities in nineteenth-century South Africa.

The social history of sport is in its late infancy. It is hoped that the contributions in this volume will help to persuade others to assist in its progress towards a sturdy academic childhood.

PART ONE

CULTURE AND SPORT IN BRITAIN

1

Sport and Industrialization: An Economic Interpretation of the Changes in Popular Sport in Nineteenth-Century England

Wray Vamplew

I

In the mid-eighteenth century the English economy began an accelerated structural transformation in which industry came more into prominence. In turn, this industrialization had an impact on popular sport as it called for new work patterns. Before the Industrial Revolution, when agriculture dominated the economy, work was much more intermittent than it was to become and there was substantial recreational time available to the working man.[1] This discretionary leisure time was reduced with industrialization. Admittedly, hard and sustained effort had been demanded by agriculture at harvest time, but in industry, particularly factory industry, the pressure was less seasonal and more unrelenting. What was required in the factories was regular hours and, above all, long hours. Unit overhead costs could only be reduced if the growing volume of machinery and other capital equipment was intensively employed, particularly as in some industries the initial productivity gains of the new methods of production over the old were not significantly large.[2] Even in relatively unmechanized industries economies could be obtained by task specialization and the division of labour. In such situations there was much to be gained from the synchronization of labour with all members of a production team working the same hours.

It was no easy task to persuade a labour force to accept the new work disciplines.[3] Not only was working from dawn to dusk for six days a week virtually throughout the year alien to traditional work patterns, but also the location of industry meant that, for many, industrial employment would involve not only occupational but also geographical mobility. Hence, in the early stages of factory industry, employers faced a labour shortage which lessened the effectiveness of disciplinary measures such as fines or dismissals. Even where successful, such methods served only to obtain a minimum performance: they did not change the labour force's attitude to work. A possible longer-term solution lay in the increased earnings which could be obtained for working longer hours. Certainly the cash stimulus appeared to work for agricultural labourers at harvest time when long hours of unremitting toil would be put in, but this lasted for only a few weeks and, to the labouring poor, could make all the difference between basic subsistence and a more comfortable living. Whether it would work all the year round was less clear, especially as many contemporaries suggested the existence of a backward sloping supply curve of labour, as epitomized in Arthur Young's famous dictum that 'every one but an idiot knows that the lower classes must be kept poor or they will never be industrious'.[4] Yet the supply of labour problem was perhaps really a supply of goods problem. Unless the workers' consumption horizons could be stretched, the labour force would exhibit a preference for leisure which would lead to an output below that desired by their industrial employers. An increased range of consumer goods was vital as a demonstration of the better material life which hard work could bring. Increased imports, the emergence of the products of the Industrial Revolution, and the shop window effect of the company store served to instill the consumption habit in the work-force. Once the workers had accepted that they wanted higher consumption standards, not only were wage incentives rendered more effective, but the way was open for the use of fines and dismissals as a disciplinary threat to those standards.

The acceptance by the work-force that leisure cost money enabled working hours to be extended, but inevitably there was a limit to the trade-off between less leisure and more income. Even for those workers with low leisure preference, Sundays were generally free from labour and, despite increasing pressure, numerous holidays were still recognized. What their workers did in this free time was of concern to many industrial and commercial employers, since certain leisure activities could be detrimental to productivity: injuries from violent sports could keep workers away from their desks or their machines; excessive drinking could result in absenteeism and hangovers; and gambling could undermine the work ethic. Employers thus supported the more general middle-class move to impose its own morality on a society in which the 'aristocracy and working class were united in their drunkenness, profaneness, sexual indulgence, gambling, and love of cruel sports'.[5] The traditional way of life was to be discredited and the importance of work upgraded: the idle rich and the idle poor were both to be condemned. Respectability was to be the aim of all society. Hence a campaign was mounted against drink, idleness, bad language, uncleanliness, and popular customs, holidays and sports.

Economic change was not confined to the industrial sector: agriculture too underwent a transformation. More land was brought into cultivation, yields were increased, fodder crops reduced the autumnal slaughter of live-stock, legumes replenished the fertility of the soil, enclosure facilitated experimentation and the development of mixed farming enabled arable and pastoral farming to advance together.[6] Such changes meant that during the eighteenth century the English population became better fed, healthier and more energetic. Agricultural change thus promoted the possibility of an altered work/leisure relationship, particularly in those sectors of the economy where regular intensive labour was required.[7] Moreover, even though climate and calendar remained the major determinants of agrarian work patterns, the changes also influenced working habits on the land. Although rural population expansion lessened the incentive to mechanize, there was still substantial investment in agriculture via enclosure, consolidation, road building, fencing, drainage and the construction of farm buildings. One estimate is that fixed capital formation in agriculture rose from £2.18 million per annum in the period 1761–70 to £4.06 million per annum in 1801–10, and that working capital rose from £138 million in 1760 to £185 million in 1800.[8] Although they were perhaps less profit-maximizing than industrialists because of land's value as a social and political asset, this investment did make landowners and larger farmers more concerned with farm productivity and the optimum use of resources. Their position as employers was strengthened by the over-supply of farm labour due both to population expansion and to the financial undermining of small-scale owner-occupiers by the costs of enclosure;[9] this enabled the landowners and larger farmers to demand, if not a daily increase in hours, then certainly an annual one.

Agricultural change also directly influenced the nature of rural recreation. Under the growing influence of the market economy, custom was giving way to an emphasis on private property rights. Not only were the customary economic practices of gleaning and fuel-gathering becoming less tolerated, but so was the traditional 'right' to play sport in a field at certain times of the year.[10] The open space available for recreation was further reduced by the loss of common rights consequent upon enclosure. Finally, fodder crops and other improvements had given man some control over his environment and weakened the rationale for the Maypole and other fertility rites which originally had been central to many rural holidays.[11]

Thus, both in the agricultural, and particularly in the industrial, sectors, there were strong economic influences acting to change traditional popular sports. Moreover, workers were generally unorganized and frequently unable to resist their employers' demands, especially as population growth had swung bargaining power in favour of capital. These economic forces were supplemented by the efforts of evangelical reformers who believed that brutal sports led directly to moral and spiritual disorders, by pressure groups such as the Royal Society for the Prevention of Cruelty to Animals (RSPCA) and the Lord's Day Observance Society which campaigned against some animal sports and against sporting on the Sabbath and by those who feared that mob football could turn easily into the riot which it resembled, with consequent damage to person, property and social stability.[12]

Unfortunately, the chronological impact of the forces arraigned against the traditional sports is not easy to determine. Several historians have argued that during the Industrial Revolution such sports were seriously undermined, but they tend not to be specific about dating either the cause or the effect.[13] What can be suggested is that, despite a long period of attrition, many traditional sports were still being played in the 1820s.[14] Several factors account for this resilience. A major one was the failure of the middle class to secure effective political power before the 1830s. Only then did Parliament legislate against animal-baiting and cock-fighting and municipal bye-laws were passed to outlaw street football.[15] Prime responsibility, however, lay with the economy. It is becoming apparent that the Industrial Revolution was a slower and much more drawn-out process than used to be believed.[16] Industrial workers remained a distinct minority of the occupied population until the middle of the nineteenth century and, despite the mushroom growth of some cities, the majority of the population could be classed as rural inhabitants until that time.[17] Clearly in 1830 England was far from being an urbanized industrial nation and thus could accommodate some of the old sporting activities.

Moreover, although the urbanization which had occurred provided sports entrepreneurs with concentrated markets, outside London, where the vast populace had long stimulated commercialized leisure for a mass market,[18] no sports events for the working-class spectator were held on a regular basis. Race meetings remained associated with annual holidays, prize-fights continued as spasmodic arrangements and quality provincial cricket matches were still infrequent occurrences.[19] The basic reason for this was insufficient working-class spending power. Although *industrial* workers made real income gains from around 1820,[20] the structural shift within the economy by 1830 was insufficient to allow the working man to afford more than an occasional outing. Regular sports spectatorship required further economic development and larger increases in discretionary spending power.

II

From 1830 the sports scenario began to change significantly. On the one hand, a shift in the locus of political power led to the national outlawing of brutal animal sports and a clampdown by many local authorities on football, pugilism, and other traditional violent sports. On the other, the spread of the railways and a rise in working-class spending power pointed the way to future developments.

Although the railways granted a temporary reprieve to some of the oppressed traditional sports, in particular prize-fighting, by allowing participants and spectators to travel to isolated areas away from magisterial interference, they actually revolutionized sport in England by widening the catchment area for spectators and by enabling participants to compete nationally. Railway specials to selected sports events soon joined the seaside excursions as part of working-class leisure.[21] Two sports especially benefited: cricket and horse racing. The late 1840s and 1850s witnessed the development of touring professional teams, beginning with William Clarke's first All-England XI, which attracted large crowds wherever they played.[22] The railways made a major breakthrough in the transport of horses and, together with the easier access which they provided for racegoers, led to racing becoming a genuine national sport rather than one pursued only at a local or regional level.[23]

An additional stimulus to such spectator sport was that more of the population became able to afford to pay for their leisure activities. All commentators agree that after the 1840s there was improvement in real wages.[24] Admittedly, neither race-meetings nor prize-fights charged entrance fees to spectators, apart from those desiring selective vantage points, but other complementary expenses were incurred such as travel, betting and refreshment. Higher earnings meant that more workers could finance such spending more frequently.

As the fruits of industrialization were passed on to the working class it became possible for them to purchase, through improved diet and living standards, more energy. In turn this encouraged two middle-class groups to attempt to persuade workers, or their children, to participate actively in sport. First, beginning in the 1860s, and accelerating in the 1870s with the expansion of a national school system, muscular Christians took team games to the urban working class in an effort to evangelise through sport.[25] Pursuing their policy of a healthy mind in a healthy body resulted in 25 of the 112 soccer teams in Liverpool in 1885 having religious affiliations.[26] In Birmingham in the 1870s and early 1880s almost 21 per cent of cricket clubs and 25 per cent of soccer clubs also had such connections.[27] Second, some industrial employers and business proprietors saw sport as having a utilitarian function in promoting human capital formation, and thus they sponsored or assisted in the formation of works soccer and cricket teams as a means of reducing labour turnover by creating loyalty to the firm and perhaps also to increase productivity by keeping their workers fit.[28] Unfortunately the chronology of this involvement is difficult to determine. Pilkingtons, the glass-makers, were running a cricket team in the 1860s, but most nineteenth-century works teams appear to have been a product of the last two to three decades.[29]

Nevertheless, despite the vast increase in the supply of energy in the late nineteenth century, particularly through the imports of cheap meat and grain, it does not seem that the great majority of working men were stimulated to more active sports participation. Detailed studies of both Birmingham and Liverpool have outlined the growth in provision of sports facilities and of parks and open spaces where games could be played. They have also shown the development of league and cup competitions in a variety of sports.[30] However, quantitative data on the numbers actively involved in sport are not easy to find. The sole national figures to hand are Football Association estimates of between 300,000 and 500,000 amateur soccer players around 1910.[31] As the total male population aged between 15 and 39 totalled some seven and a quarter million in 1911, this can hardly be seen as representing a substantial commitment to sports participation. A similar conclusion can be drawn from the city studies mentioned above. Liverpool had 224 cricket teams in 1890 and 212 soccer clubs in 1892, whereas its total population in 1891 was 518,000, and Birmingham possessed 214 cricket clubs in 1880 though its population a year later was 401,000.[32] Even allowing for female non-involvement and several teams to a club, the figures still suggest that playing sport, at least at an organized level, was very much a minority activity.

III

What is clearer is that the late nineteenth century witnessed a significant increase in sports spectatorship, particularly at gate-money events. Large crowds at sports fixtures were nothing new, but now large crowds were being attracted regularly. By the 1890s race crowds of 10,000 to 15,000 were not unusual; double this could be expected at leading meetings, and perhaps 70,000 to 80,000 at a major Bank Holiday event.[33] In soccer it became 'no rare thing in the North and Midlands for twenty to thirty thousand people to pay money to witness a League match or important cup-tie'.[34] Indeed commercialized spectator sport for the mass market became one of the economic success stories of late Victorian England. All over the nation, and in a wide variety of sports, entrepreneurs and sports club executives enclosed grounds, built stadia and charged gate-money. The growth of sports professionalism was also stimulated. By 1910 there were over 200 first-class professional cricketers plus several hundred county ground staff and league professionals; some 400 jockeys and apprentices were seeking rides in horse races; and in soccer there were 6,800 registered professionals, many of them part-timers but still earning money from playing sport.[35] Leading professionals could obtain high economic rewards. The benefit match and subscription lists for George Hirst, the Yorkshire Test cricketer, yielded him the present-day equivalent of £100,000; even more impressive was the £250,000 annual earnings of champion jockey Fred Archer.[36]

Changes in economic variables played a major role in this development. Foremost was the substantial increase in working-class spending power. There are, of course, difficulties in generalizing about *the* working class: at any time the skilled artisan probably could afford a wider range of recreational activity than the labourer or factory hand. Nevertheless, most economic historians would accept that the bulk of the working class experienced a rise in real incomes between the 1850s and the end of the century; this was estimated on average at some 70 per cent.[37] Simultaneously, increased life expentancy both deepened and widened the potential market for spectator sport.[38] Moreover, between 1851 and 1901 the proportion of the population living in urban areas rose from 50.2 per cent to 77.0 per cent[39] which would have produced concentrated markets for recreational entrepreneurs, though perhaps there was a lessened need for absolute population density because of developments in transport technology, particularly railways, and, to a lesser extent, tramways. A more important factor was the widespread adoption of Saturday afternoon as a time free from work. Unlike most other goods, sport has a temporal aspect to its demand in that generally it is consumed at particular points in time. Thus the time available for sports consumption is not just a matter of the volume of 'free time' but also where these non-working hours are located in the work/leisure calendar. Factory legislation and trade-union pressure established the free Saturday afternoon as a norm in many trades by the 1870s, though unskilled labour in several areas had to wait another decade or so to join their more fortunate brethren.[40] Since electrical technology was not yet capable of efficiently illuminating night games and Sunday sport was still taboo, Saturday afternoon was an ideal time-slot for spectator sport.

Entrepreneurs and others responded to the stimulus. As the economic benefits of industrialization were passed down to the working class, sport and other leisure activities became industries in their own right. Heavy investment was undertaken in ground facilities. Raising the required capital forced many sports clubs to adopt company ·status with shareholders and limited liability. The covering of these and other costs necessitated the holding of events on a regular basis and, to ensure that these could go ahead, advantage was taken of modern technology in the form of drainage and stand construction to combat the English climate. To attract larger crowds, the organization of events was improved, particularly regarding running to time. Crowd control measures were adopted, leagues were introduced in team sports to ensure competitive fixtures and arrangements were made with tram and railway companies to transport spectators to and from the events. As sport became more commercialized its character and structure were often changed in the interests of Mammon. The accepted rules of the game were tampered with and established sports were altered in attempts to get more customers through the turnstiles. In horse racing, for example, long-distance staying events gave way to sprints, two-year-old and handicap racing, all of which had a sufficient degree of unpredictability about the result to attract working-class punters.[41] In rugby, the split between the amateurs and the professionals produced two distinct versions of the game with differences in the number of players, methods of scoring and style of play.[42] In soccer, clubs from less densely populated areas were sometimes voted out of the Football League because they could not produce enough revenue at the gate.[43]

IV

Although money was being made out of man at play, it can be suggested that, despite the emphasis on private rather than public funding, profit-maximization was not necessarily the aim of those who invested in the sports industry. At one end of the spectrum of motivations were those who sought direct profits such as boxing promoters and some owners of thoroughbred racing, and especially breeding bloodstock. Yet even in this category there were rentiers satisfied with the safe return rather than the high-risk, high-yield investment: such a group would include the shareholders who financed the Edgbaston cricket ground and felt that they were entitled to a fair return though their main objective was 'not to make dividends but to advance the interests and position of the county club'.[44]

Others sought profits more indirectly. As well as the businessmen who patronized works teams as part of their industrial welfare policies, there were the penny capitalists and wealthier organizers of illegal off-course betting,[45] cycle manufacturers who regarded the sponsoring of meetings and riders as a form of advertising[46] and those builders, caterers and sports outfitters who took shares in football clubs in the expectation of contracts.[47] Another major group of football shareholders who looked to benefit indirectly was the drink trade who hoped that fans would celebrate their victories or drown their sorrows in public houses owned by men who supported their club financially.[48] Transport enterprises expected to benefit from sports passenger traffic and hence extended facilities to football grounds and race-tracks, occasionally also sponsoring races at the latter.[49] Another example of a symbiotic relationship was that with the press which provided free publicity for sports events and in turn found ready customers among sports fans.[50]

A third category of sports investor seems to have been either uninterested in profits or gave priority to other objectives. Indeed there is a major debate among sports economists as to whether gate-money sports clubs were profit or utility maximizers, for the two objectives cannot be attained simultaneously.[51] Historical research is suggesting that in England, although finance was important simply to carry on, few cricket or soccer clubs put profits first, but more detailed work is required on the decision-makers in these clubs before specific motivations can be attributed.[52] Examples of uneconomic (or rather non-economic) investment in sport can be found at all levels of society. Some upper-class and middle-class expenditure can be viewed as consumption, sometimes of the conspicuous variety despite its apparent investment nature. The net cost per kilogram of some game birds must have been high considering the establishment, maintenance and killing expenses.[53] Clearly, as in racehorse ownership, there was a strong social status element to the heavy expenditure involved.[54] Other members of the middle class laid out substantial sums to purchase social exclusivity in their capital-intensive and land-extensive golf clubs.[55] Yet others were prepared to subsidize their local or county cricket team for reasons of civic pride, county allegience or national jingoism with no thought of the sport's being a commodity in the market place.[56] Some of these may have been among that group who sought to use sport as a means of social control of the lower orders by investing in municipal facilities and sponsoring teams.[57] Even some of the working class were prepared to purchase a few shares in their local soccer club with no regard to financial return: to them shareholding was simply an extension of being a fan.[58]

Unconventional economic practices also occurred in the labour market for professional sportspersons. In its discrimination against women the market was not atypical,[59] but in other respects, with its severe limitations on mobility and earning power, the sports labourer operated in a market very different from that of other occupations. Indeed many professional sportsmen were virtually bonded men. For example, once soccer players had signed for a club they were no longer free to select their employer, and could only change clubs if a potential employer was willing to pay the transfer fee demanded by the existing one. Adding to the injustice was the imposition, in 1901, of a maximum wage which in some instances halved the earnings of star players.[60] Professional cricketers were free to change county clubs but had to stand out of the first-class game for two years.[61] Despite such draconian restrictions there was little unionization or overt labour unrest among professional sportsmen, surprising perhaps in a period which witnessed unprecedented unionism and strikes among other working men.[62] Indeed the Association Football Players' Union was the only trade union established for professional sportsmen before 1914 and, in 1910, it attracted only 500 members, approximately seven per cent of the professionals registered with the Football Association.[63] Several factors militated against sports unionism, including the competitive nature of an occupation which always has to have winners and losers. Given the brevity of most careers in sport,[64] too many professionals were out to make what they could from the game to spare time for collective action, save perhaps at club level where there would be relatively easy communication between players. In some sports even losers had low opportunity costs which could have lessened the incentive to band together for higher wages, except in soccer where restrictions prevented the stars from earning their economic rents. Perhaps for many professionals there was substantial job satisfaction, and the adulation of the crowd could provide psychic income which, if coupled with an emotional attachment to the game and loyalty to one's club, could undermine the influences pushing for unionism. Finally, in most sports, employers' organizations had pre-empted the provision of welfare schemes which again weakened the move towards unionism.[65]

Another unusual feature of the sports market is that generally at least two firms (players or clubs) are required to manufacture the sports product, so that, whereas the ideal market position of a conventional business would be monopoly, this is less desirable in sport, for a champion without challengers has nothing to sell. Yet it is not within the province of any individual club deliberately to eschew becoming dominant: it must always play to win. Any policy aimed at the prevention of a monopoly must be imposed on the clubs by some central body. This occurred, to a greater or lesser degree, in all gate-money team sports in the late nineteenth century, generally by means of regulations designed to prevent the wealthiest clubs from recruiting all the best talent. When such rules were successfully applied they served to produce a greater equality of playing competition, a feature which most sports economists believe, other things being equal, will attract larger crowds than one-sided contests.[66] Thus it is possible that the regulations were part of the arrangements of a profit-maximizing cartel, but, in the light of the profit versus utility debate, it may be that they were designed more to ensure the survival of the relevant leagues; for if one team was overwhelmingly dominant then poor gates might render the weaker teams unviable and cause the league to collapse. Adding strength to this argument, as far as soccer is concerned, is that the Football Association also interfered with the rights of capital and limited dividends to a maximum of 5 per cent,[67] thus reducing the chances that high profits for distribution were a major objective of the clubs.

V

During the nineteenth century popular sport in England was influenced by industrialization in three overlapping phases. First, the work demands of the Industrial Revolution produced changes both in the pattern of work and leisure, initially increasing the former at the expense of the latter, and in the character of many sports, though the extent of the change by 1830, the end of the 'classical' Industrial Revolution, should not be exaggerated. Second, from the middle of the century some industrialists began to appreciate that sport could have a utilitarian function and thus began to patronize works teams. Finally, structural change and productivity-raising innovation associated with industrialization eventually led to both increased real incomes and leisure time which, in turn, stimulated a demand for commercialized spectator sport. Sport, then, underwent its own revolution and became an industry in its own right, though not all sports firms or entrepreneurs acted as in conventional business, many instead putting performance on the field of play ahead of profits.

18   BRITISH CULTURE AND SPORT AT HOME AND ABROAD, 1700–1914

NOTES

1. Pre-industrial work patterns are discussed in W. Vamplew, 'The Influence of Economic Change on Popular Sport in Britain 1600–1900' in R. Crawford, *Proceedings of the First Australian Symposium on the History and Philosophy of Physical Education and Sport* (Melbourne, 1980), pp. 126–32.
2. J.S. Cohen, 'The Achievements of Economic History: The Marxist School', *Journal of Economic History* XXXVIII (1978), 52–3; J.S. Lyons, 'The Lancashire Cotton Industry and the Introduction of the Powerloom, 1815–1850', *Econ. Hist.* XXXVIII (1978), 284. It was not until the widespread adoption of steam power in the mid-nineteenth century that productivity increased sufficiently to allow a general reduction in industrial working hours. C.N. Von Tunzelman, 'Technological Progress During the Industrial Revolution' in R. Floud and D.N. McClosky, *The Economic History of Britain Since 1700* (Cambridge, 1981), p. 158.
3. Later works have only slightly modified the view presented in the seminal article by S. Pollard, 'Factory Discipline in the Industrial Revolution', *Economic History Review* XVI (1963/64), 254–71.
4. Quoted in D.C. Coleman, 'Labour in the English Economy of the Seventeenth Century', *Econ. Hist. Rev.* VIII (1956), 280.
5. H. Perkin, *The Origins of Modern English Society 1780–1880* (London, 1972 ed.), p. 277.
6. E.L. Jones, *Agriculture and the Industrial Revolution* (London, 1974).
7. Although the agrarian nature of the pre-industrial economy determined the seasonal nature of productive activity, Freudenberger and Cummings ('Health, Work and Leisure Before the Industrial Revolution', *Explorations in Economic History* 13 (1976), 1–12) have argued that the long periods of less intensive work at other times were a function of the population's physical capacity to work. Low and unpredictable food supplies combined with debilitating diseases to render energy a scarce resource and to necessitate long stretches of rest and recuperation after prolonged bouts of hard work. Leisure preference was thus seen as involuntary in that idleness was essential to health.
8. C.H. Feinstein, 'Capital Formation in Great Britain' in P. Mathias and M.M. Postan, *The Cambridge Economic History of Europe* (Cambridge, 1978) Vol. 7, part 1. All figures are in 1851–60 prices.
9. M.E. Turner, *Enclosures in Britain 1750–1830* (London, 1984), pp. 64–81.
10. R.W. Malcolmson, *Popular Recreations in English Society 1700–1850* (Cambridge, 1973), pp. 111–16.
11. A. Howkins, *Whitsun in Nineteenth Century Oxfordshire* (Oxford, 1973), pp. 1–2.
12. See, for example, B. Harrison, 'Religion and Recreation in Nineteenth Century England', *Past and Present* 38 (1967), 98–125; Malcolmson, op. cit., 106–7; R. Burt, *Industry and Society in the South West* (Exeter, 1970) pp. 71–106.
13. A recent survey by W.J. Baker in 'The Leisure Revolution in Victorian England; A Review of Recent Literature', *Journal of Sport History* 6 (1979), 76–87, has the sports under successful attack from the second half of the eighteenth century to beyond 1835.
14. Based on a survey of P. Egan, *Book of Sports* (London, 1832), W. Hone, *The Everyday Book*, 3 vols. (London, 1830), and *The Sporting Repository* (London, 1822). More local studies are required to help date the decline of such sports and to assess whether or not this was associated with industrialization and/or urbanization.
15. 5 and 6 William IV, c 59; A. Delves, 'Popular Recreation and Social Conflict in Derby, 1800–1859' in E. and S. Yeo, *Popular Culture and Class Conflict 1590–1914* (Brighton, 1981), pp. 89–127.
16. The most recent evidence has been supplied by C.K. Harley, 'British Industrialisation Before 1841: Evidence of Slower Growth During the Industrial Revolution', *J. Econ. Hist.* XLII (1982), 267–90; J.G. Williamson, 'Why was British Growth So Slow During the Industrial Revolution?', *J. Econ. Hist.* XLIV (1984), 687–712; and N.E.R. Crafts, 'British Economic Growth, 1700–1831: A Review of the Evidence', *Econ. Hist. Rev.* XXXVI (1983), 177–99.

17. C.H. Lee, *British Regional Employment Statistics 1841–1971* (Cambridge, 1979); J. Saville, *Rural Depopulation in England and Wales 1851–1951* (London, 1957), p. 61.
18. D. Brailsford, *Sport and Society* (London, 1967), pp. 215–17.
19. W. Vamplew, *The Turf* (London, 1976), p. 25; P. Egan, op. cit., *passim*; C. Brookes, *English Cricket* (Newton Abbot, 1978), pp. 81–100; W. Vamplew, 'Sport' in R.J. Morris and J. Langton, *Atlas of the Industrial Revolution in Britain* (London, 1986), pp. 198–201.
20. P.H. Lindhert and J.G. Williamson, 'English Workers' Living Standards During the Industrial Revolution: A New Look', *Econ. Hist. Rev.* XXXVI (1983), 11.
21. D.A. Reid, ' "To Boldly Go Where No Worker Had Gone Before": The Social and Economic Significance of the Railway Excursion, 1846–1876', paper presented at Eighth International Economic History Congress, Budapest, 1982.
22. Brookes, op. cit., pp. 101–17.
23. Vamplew (1976), pp. 29–37.
24. For a bibliography of the literature see Lindhert and Williamson, loc. cit., n. 1.
25. J. Walvin, *Leisure and Society 1830–1950* (London, 1978), pp. 86–7. A. Mason, *Association Football and English Society 1863–1915* (Brighton, 1980), pp. 12–13.
26. Walvin, op. cit., p. 87.
27. Mason, op. cit., p. 26.
28. Ibid., pp. 28–30. More research is required before motivations can be confidently attributed.
29. Conclusion based on a survey of 100 business histories, though 67 of these had no reference to recreation. T.C. Barker, *The Glassmakers* (London, 1977), p. 93.
30. D.D. Molyneux, 'The Development of Physical Recreation in the Birmingham District from 1871–1892', M.A., University of Birmingham (1957); E.H. Roberts, 'A Study of the Growth of the Provision of Public Facilities for Leisure Time Occupations by Local Authorities of Merseyside', M.A., University of Liverpool (1933); R. Rees, 'The Development of Physical Recreation in Liverpool During the Nineteenth Century', M.A., University of Liverpool (1968).
31. G. Green, *The History of the Football Association* (London, 1953), p. 251, p. 261.
32. Molyneux, op. cit., pp. 26–7; Rees, op. cit., p. 69; B.R. Mitchell and P. Deane, *Abstract of British Historical Statistics* (Cambridge, 1962), p. 6, pp. 24–5.
33. C. Richardson, *The English Turf* (London, 1901), p. 213.
34. W.J. Oakley and M. Shearman in M. Shearman, *Football* (London, 1895), p. 166.
35. Figures derived from *Wisden's Cricketers' Almanack*, *The Racing Calendar*, and *Football Association Circular on Financial Arrangements Between Clubs and Players* 10 January 1910.
36. *Yorkshire County Cricket Club Year Book* (1905), p. 11; J. Welcome, *Fred Archer: His Life and Times* (London, 1967), p. 88.
37. Calculation based on Mitchell and Deane, op. cit., pp. 343–4. For a discussion on the standard of living in the late nineteenth century see T.R. Gourvish, 'The Standard of Living 1890–1914' in A. O'Day, *The Edwardian Age*.
38. R. Mitchison, *British Population Change Since 1860* (London, 1977), pp. 39–57.
39. Saville, op. cit., p. 61.
40. Walvin, op. cit., p. 60; D.A. Reid, 'The Decline of Saint Monday 1766–1876', *Past and Present* 71 (1976), 76–101; E. Hopkins, 'Working Hours and Conditions During the Industrial Revolution: A Re-Appraisal', *Econ. Hist. Rev.* XXXV (1982), 52–66.
41. Vamplew (1976), op. cit., pp. 38–48.
42. E. Dunning and K. Sheard, *Barbarians, Gentlemen, and Players* (Oxford, 1979).
43. What soccer enthusiast today can recall Ardwick, Bootle, Burton Swifts, Darwen and Northwich, all foundation members of the Football League second division?
44. Quoted in L. Duckworth, *The Story of Warwickshire Cricket* (London, 1974), p. 19. The actual dividend paid turned out to be 6 per cent per annum. *Annual Reports of Warwickshire Cricket Ground Company 1907–14.*
45. J. Benson, *The Penny Capitalists* (Dublin, 1983), pp. 70–73; Vamplew (1976), op. cit., pp. 207–12.

46. P. Bilsborough, 'The Development of Sport in Glasgow, 1850–1914', M. Litt thesis, University of Stirling, 1983, pp. 217–18, shows this to be the case in Scotland. Presumably it also occurred in the larger markets south of the border.
47. Mason, op. cit., Ch. 2; S. Tischler, *Footballers and Businessmen* (New York, 1981), Ch. 4.
48. W. Vamplew, 'Borderline Differences: A Comparative Analysis of Shareholders in English and Scottish Football Before 1914', *Flinders Occasional Paper in Economic History* No. 2 (1984). See, for example, prospectuses of English football clubs held at Companies Registry, Cardiff.
50. Mason op. cit., pp. 187–95; Vamplew (1976), op. cit., pp. 220–23; Walvin, op. cit., p. 90.
51. For bibliographical references see P.J. Sloane, *Sport in the Market* (London, 1980) and W. Vamplew, 'The Economics of a Sports Industry: Scottish Gate-Money Football, 1890–1914', *Econ. Hist. Rev.* XXXV (1982), 550.
52. W. Vamplew, 'Profit or Utility Maximisation? An Analysis of English County Cricket Before 1914' in W. Vamplew, *The Economic History of Leisure: Papers Presented at the Eighth International Economic History Congress* (Budapest, 1982); S.G. Jones, 'The Economic Aspects of Association Football in England, 1918–39', *British Journal of Sports History* 1 (1984), 286–99. However, Tischler, op. cit., especially Ch. 3, would dissent from this view.
53. See the many articles in the *Badminton Magazine* 1895–1914 and Lord Walsingham and R. Payne-Gallway, *Shooting* (London, 1920).
54. Vamplew (1976), op. cit., Ch. 11.
55. J.R. Lowerson, 'Joint-Stock Companies, Capital Formation and Suburban Leisure in England, 1880–1914' in Vamplew (1982), op. cit., pp. 61–71; J.R. Lowerson, 'Middle Class Sport, 1880–1914' in *Social History of Nineteenth Century Sport: Proceedings of the Inaugural Conference of the British Society for Sports History* (University of Liverpool, 1982).
56. K.A.P. Sandiford, 'Cricket and the Victorian Society', *Journal of Social History* 17 (1983), 303–17.
57. For examples see P. Bailey, *Leisure and Class in Victorian England* (London, 1978), H. Cunningham, *Leisure in the Industrial Revolution* (London, 1980), H.E. Meller, *Leisure and the Changing City 1870–1914* (London, 1976).
58. Vamplew (1984), op. cit.
59. The extent of discrimination is difficult to determine. With American jockeys, Scottish footballers, and women in all sports, there may have been perceived differences in skill, but this is less applicable to the amateur cricketers who were able to oust professionals from many county teams during the university vacations.
60. Tischler, op. cit., pp. 89–104; W. Vamplew, 'Playing for Pay: The Earnings of Professional Sportsmen in England 1870–1914' in R. Cashman and M. McKernan, *Sport, Money, Morality and the Media* (Sydney, 1981), pp. 118–26.
61. *Wisden's Cricketers' Almanack* (1887), p. 308.
62. E.H. Hunt, *British Labour History 1815–1914* (London, 1981), pp. 295–338; S. Meacham, 'The Sense of an Impending Clash: English Working Class Unrest Before the First World War', *American Historical Review* 77 (1972), 1343–64.
63. Calculated from *Subscription Register of Association Football Players' Union 1910; Football Association Circular on Financial Arrangements Between Clubs and Players*, 10 January 1910.
64. W. Vamplew, 'Close of Play: Career Termination in English Professional Sport 1870–1914', *Canadian Journal of History of Sport* XV (1984), 64–79.
65. This is considered more fully in W. Vamplew, 'Not Playing the Game: Unionism in British Professional Sport 1870–1914', *Brit. J. Sports Hist.* 2 (1985), 232–47.
66. For example Sloane, op. cit., 25; R. Noll, 'Attendance and Price-Setting' in Noll, *Government and the Sports Business* (Washington, 1974), p. 156.
67. Green, op. cit., p. 153.

2

Bonaparte and the Squire: Chauvinism, Virility and Sport in the Period of the French Wars

Derek Birley

I knew no harm of Bonaparte and plenty of the Squire,
And for to fight the Frenchman I did not much desire;
But I did bash their baggonets because they came array'd
To straighten out the crooked road an English drunkard made.

G.K. Chesterton, 'The Rolling English Road'.

Earlier wars had been about religion and territory; this one had its roots in class. Many in England had high hopes of the French Revolution. Tom Paine's *The Rights of Man* sold nearly 200,000 copies. Pitt's misgivings about the war with France did not save him from accusations of 'Sending troops to be swamped where they can't draw their breath, and buying a fresh load of taxes with death.' Handbills protested at 'this vile and wicked war' and 'the oppressor of the people that reigneth upon the throne'.[1]

The Royal Family and the aristocracy had a vested interest in stemming the tide of Jacobinism, but the common people were more concerned with the price of bread. The harvest of 1793 was a bad one, and food riots broke out spasmodically. When the King drove to open Parliament that autumn, sections of the crowd yelled 'No Pitt, no war – bread, bread, peace, peace' and 'Down with George'. For the lower orders, furthermore, the war brought curtailed freedom as well as heavy taxation. Habeas Corpus was an early casualty; the Combination Acts outlawed trades unions and steered them towards sedition.

Farmers and landowners did well out of war-time conditions: farm labourers got some increase in wages but in most districts this was insufficient to match the rise in prices. In the north, competition for labour from industry helped keep up agricultural wages, but in much of the south, county magistrates shirked fixing a minimum wage and directed Poor Law authorities to make up deficiencies to bare subsistence level. Despite stark poverty, few could be persuaded to enter the armed forces voluntarily. In April 1795 *The Times* reported that 'Mendoza and Bill Ward, finding the blackguard exercise of boxing has fallen into disrepute took up the gentle art of crimping and became acting sergeants at a house in St George's Fields'. Conditions in the army were improved sufficiently to stave off incipient mutiny: the navy, relying on the methods (and the pay) that had served it for a century, was less fortunate. Mutinies at the Nore and Spithead were in effect strikes against poor conditions. They eventually brought some improvement but the ring-leader was hanged and the episode strengthened the conviction of the better sort that Jacobinism was the enemy at home as well as abroad.

Most persons of rank, unless they were professionally involved in politics or the forces, felt no urge to take the fight to the French. Pitt himself hoped for a quick and relatively painless end to it through the traditional, profitable British method of fighting the war at sea and in the colonies while funding Britain's allies to fight continental land battles. The navy was a splendid protective screen: as Trevelyan put it '. . . the war was in the newspapers but it scarcely entered the lives of the enjoying classes. . . While Napoleon was ramping over Europe the extravagance and eccentricity of our dandies reached their highest point in the days of Beau Brummell. . . .'.[2]

Sport was among the principal amusements of the 'enjoying classes', and it often led to a conspicuous display of wealth that invited radicalism. Yet it arguably helped to avert social revolution and enabled Britain to fight the war. This is not to suggest that it fostered the Newbolt concept of leadership. Whatever was bred on the playing fields of Eton it was not team spirit. Nor should we accept uncritically Trevelyan's dictum that 'If the French noblesse had been capable of playing cricket with their peasants their châteaux would never have been burnt'.[3] The match for a thousand guineas in 1794 between the Earl of Winchelsea and Lord Darnley, old Etonians and members of the fashionable Marylebone Cricket Club, had more to do with metropolitan frivolity than rural harmony. Lord Frederick Beauclerk, the notably unchristian vicar of St Albans, who was the leading cricketer of the day, looked to cricket to augment his income and was notorious for sharp practice.

Indeed, on the face of it, sport was an obstacle rather than an aid to social cohesion. Sporting gatherings, especially those which attracted the lower orders, had long been regarded by successive governments as potential fronts for subversive activity, and this was a time when the enclosure movement had sparked off lawless behaviour, sometimes with political overtones. The Court of Common Pleas heard many charges of breaking into closes to play cricket and football. Sport apart, clubs of any kind were regarded with suspicion if their social standing was not impeccable, and some, like the London

Corresponding Society, ostensibly a working man's discussion group, were outlawed.

The more obvious problems were urban. Football hooliganism – in the streets – had plagued the London authorities since the twelfth century, and the old mass festival games had become a byword for ferocity. At Derby in 1796 a football player was drowned in the Derwent, the inquest jury denounced the sport as 'disgraceful to humanity and civilisation' and the Mayor and Justices 'unanimously resolved THAT SUCH CUSTOMS SHALL FROM HENCE-FORTH BE DISCONTINUED'.[4] Even the newer, Italianate game could arouse fierce passions. A match in 1793 between 'six young men of Norton dressed in green and six young men of Sheffield' ended in uproar with the Derbyshire men cutting off the Sheffielders' pigtails and many injuries occurring among the spectators.[5]

But in this still predominantly agricultural nation the countryside was by no means free from social and sporting tensions, even in cricket. There was, for instance, a curious episode in the Hampshire village of Hambledon, which had been the centre of first-class cricket in the 1780s. The Hambledon Club, though no longer supreme on the field of play, was still active socially until 1795. Inflation was having an effect, and the cost of the annual dinner went up from one shilling to two shillings; nevertheless there were 49 members present and they spent fourteen shillings a head on wine. The following year came a dramatic change. As the minutes of a meeting on 29 August record there were only three members present, but there were twelve guests, including 'Mr Thos Paine, "Author of the Rights of Man" '. It seems inconceivable that Citizen Paine, forbidden to enter the country and not long out of a French gaol, would have smuggled himself in simply to discuss cricket, despite his youthful interest in the game, or that his notorious book would have been mentioned except for a purpose. The Honorary Secretary, Henry Bonham, had radical leanings and it may be that his choice of friends upset the more orthodox. Be that as it may, on 21 September the terse minute 'No gentlemen present' recorded the Club's demise.[6]

However, such episodes were peripheral compared with the chronic and acute social divisions in traditional field sports. Conservation of game was a major preoccupation of landowners, and poaching reached alarming proportions as food grew dearer and scarcer. The Game Acts, together with the tithes exacted by the Church, were among the restrictions most resented by the common people, and in this war against Jacobinism there was mounting pressure in Parliament for even more severe penalties for transgressors.

II

We must look beyond sport, then, for the major reasons why revolution was averted. Economic factors apart, it seems likely that what chiefly preserved the better sort was xenophobia, the old John Bull beefeating feeling of superiority to the French that was part of popular consciousness. It helped sustain a belief in the liberty of the subject. A wartime visitor from democratic America remarked upon 'the active jealousy of liberty which exists even in the lowest orders of England'. The reality was less attractive than the myth. However, the John Bull self-image, based on the virility cult, was common to men of all classes. It had encouraged a more personalized and thus less lethal hostility between the lower orders and their betters compared with France, and in this sport played a part.

When Parliament debated a Bill to abolish bull-baiting in 1800, William Windham, Minister of War, stoutly defended the people's right to such amusements, denouncing the hypocrisy of the abolitionists who had 'a most vexatious code of laws for the protection of their own amusements'. Windham's performance allowed R.B. Sheridan to raise a laugh by referring to his 'truly Jacobinical doctrine' for Windham was no populist but an old Etonian enthusiast for the virility cult, notably prize

Boxing was a fashionable aristocratic pursuit and an important symbol of British pluck which enjoyed a boom during the war years. The prize-ring was, however, in disrepute, not least because of deaths and injuries. Its survival owed much to the patronage of the great, including the Duke of Clarence, the future William IV. In pugilism, as in racing, the other great popular sport, it was the element of gambling that brought the classes together. Gentlemen might enjoy riding or boxing but professionals from the lower classes were brought in when money was at stake.

Boxing was a rough business, socially as well as physically. The champion when the war began, Daniel Mendoza, was type-cast as a crafty Jew. His book *The Art of Boxing* emphasized the importance of attacking vulnerable parts of the body. You should hit an opponent on 'the eye brows, on the bridge of the nose, on the temple arteries, beneath the left ear, under the short ribs, or in the kidneys'. A blow on the kidneys 'deprives the person struck of his breath, occasions an instant discharge of urine, puts him in the greatest torture and renders him for some time a cripple'.[7] There was much rejoicing in 1795 when he was relieved of the title by 'Gentleman' John Jackson. However, *The Times* was not impressed: 'Yesterday a Prize Battle was fought at

Hornchurch in Essex between Mendoza the Jew and one Jackson, a publican of Gray's Inn Lane, when, as had no doubt been previously settled, the Jew appeared overpowered by the strength of the Christian'.

Indeed *The Times* disapproved of pugilism, and particularly of Jackson. 'We think it worthy of the notice of the magistracy to consider whether a man who breaks the peace should be a fit person to have a licence as a publican'. But Jackson had his sights set on higher things. He retired from active competition almost as soon as he had won the title and in 1796 set up a boxing school in Bond Street that far surpassed anything previously seen. 'Jackson's Rooms' attracted the patronage of the fashionable world and young sprigs of the nobility sparred with the great man. Windham presented Jackson with a print of a bloody Roman assassination and inscribed it with a message for 'those who are labouring to abolish what is called the brutal and ferocious practice of boxing'.

The Times also disapproved of racing, not least because of the involvement of the Prince of Wales, under the tutelage of Charles James Fox, in its machinations. The Prince had been obliged to quit Newmarket in 1791, when the dubious practices of his rider, Sam Chifney, had impelled the Senior Steward of the Jockey Club, Sir Charles Banbury, to inform the Prince that if he engaged Chifney again no gentleman would accept his wager. The Prince's withdrawal was an economic as well as a social necessity, for he was heavily in debt. By 1795, however, his unhappy but lucrative marriage allowed him a new bout of extravagance, which included raising Brighton and Lewes races to new heights of colourful splendour. Ascot was also thriving, and among the additional attractions was an early form of roulette called EO. Special booths were licensed at twelve guineas for ordinary play and forty guineas for 'the gold table'. Epsom was more accessible to, and more popular with, the public at large. The whole town became a seething mass on race days, which was good for business but brought other troubles. As a newspaper report of 1799 commented, 'The vicious and unprincipled form a tolerable proportion of the crowd and it generally follows that many atrocities are committed'.[8]

Pitt's plans for a cosy war went adrift. The Dutch withdrawal from the coalition, financial recession at home and Austria's collapse led him to make peaceful overtures in 1797. They were contemptuously rejected by the Directory in Paris. Spain changed sides and invasion seemed imminent. Naval victories staved off the immediate threat and in 1798, with Napoleon preoccupied with conquering Egypt, Nelson won a great victory at the Nile. Pitt was encouraged to form a second coalition, including Russia, Naples, Turkey and Austria, which did well on land as long as Napoleon was stuck in Egypt. After his return in 1800, however, the coalition began to crumble. Most landowners were more interested in the new Game Act which categorized groups of two or more poachers as rogues and vagabonds liable for hard labour and, on a second offence, as incorrigible rogues eligible for whipping and imprisonment or service in the armed forces.

In 1801 the antiquary Joseph Strutt produced the first social survey of sport, a distinctly hierarchical treatise. Despite Parliament's approval Strutt felt obliged to censure vulgar animal baiting. 'Bull and bear-baiting is not encouraged by persons of rank and opulence in the present day, and when practised, which rarely happens, it is attended only by the lowest and most despicable part of the people.' And even 'the sanction of high antiquity' could not gloss over the fact that cock-fighting was 'a barbarous pastime'. On the other hand hunting, hawking, fowling and fishing, along with horse racing, were lauded as 'rural exercises practised by Persons of Rank'.[9]

The same social considerations characterized Strutt's attitude to ball games. Thus he was dismissive as well as ill-informed about football: 'it was formerly much in vogue amongst the common people of England, though of late years it seems to have fallen into disrepute and is but little practised.' On the other hand, though he knew little about cricket either, he declared it a 'pleasant and manly exercise', of late years 'exceedingly fashionable, being much countenanced by the nobility and gentlemen of fortune'.

The reality was somewhat different. Metropolitan cricket matches attracted big crowds and some determined thieves. This apart, the game itself was corrupt. As one old professional confessed in later years, 'If a gentleman wanted to bet, just under the pavilion sat men ready with money down to give and take at current odds. . . Then these men would come down to the "Green Man and Still" and drink with us, and always said that those who backed us, or "the nobs" as they called them, sold the matches'.[10]

Notwithstanding its ancient origins, Strutt clearly found golf somewhat outlandish. Of the modern game he remarked only 'In the northern parts of the Kingdom, Goff is much practised. It requires much room to perform this game with propriety, and therefore I presume it is rarely to be seen at present in the vicinity of the metropolis'. This glossed over not merely sporting but social developments of a sophisticated kind. In Edinburgh the gentleman golfers of Leith had the previous year successfully petitioned the city authorities to grant them the status of 'one body politic and corporate' to be known as 'The Honourable the Edinburgh Company of

Golfers'; and their neighbours at Bruntsfield Links had formed the 'Edinburgh Burgess Golfing Society' composed largely of those engaged in trade. Strutt exuded neo-medieval London snobberies. He recalled a time over 25 years before when 'the hurling to the goals was frequently played by parties of Irishmen in the fields at the back of the British Museum . . . I have been greatly amused to see with what facility those who were skilful in the pastime would catch up the ball with the bat, and often run with it for a considerable time'.

The old, royal – and French – game of tennis retained its prestige, but its exclusiveness had long been eroded by commercialism, and it now had to endure journalistic caprice. In 1797 *The Times* reported that 'The once fashionable game of tennis is very much upon the decline. The court in the Haymarket seems to be entirely for-lorn.' (This was a punning reference to the continued addiction of Lord Lorne, later Duke of Argyll.) However, new variants such as 'fives', 'bat-fives' and 'rackets' were springing up. These were versions of the age-old handball game in which the ball was knocked up against a wall, and they were played in improvised locations such as inn-yards and, not least, the Fleet debtors' prison.

Jackson saw the commercial possibilities of using the Fives Court in St Martin's Street for boxing exhibitions. These were gloved contests during which the audience were safe from the intrusion of the law. But the real, bare-knuckle game had greater appeal, and Jem Blecher, one of a remarkable series of Bristol butchers who – along with London watermen and ethnic minorities – dominated the boxing scene, took Jackson's place as the people's champion. A relatively small man, he came to exemplify British bulldog courage against heavier opponents, and went on fighting even after he lost an eye in a rackets game.

III

The war had dwindled towards temporary truce, but it began again in 1803. Pitt, who had resigned after a disagreement with the King over Catholic emancipation, returned, though his Cabinet was weak and divided. However, a new threat of invasion was met by a surge of patriotism in which 3,000 volunteers came forward. That same year the Ellenborough Act introduced hanging for poachers offering armed resistance to arrest. So the paradox remained: Pitt the enemy of popular liberty was the standard-bearer of John Bull's ideals.

There was no invasion and thus no need for the militia to use their skills to serious purpose. The army assembled at home included many high-spirited young cavalry officers. They were prominent not only in reviving the tradition of gentlemen riders – the original jockeys – on the Flat, but also in introducing the sport of steeplechasing, pioneered in Ireland, to England. In 1804 *The Sporting Magazine* reported that in Newcastle 'a wager betwixt Captains Prescott and Tucker of the 5th Light Dragoons was determined . . . by a single horse race which we learn is denominated steeple-hunting'.

The disastrous course of the war in Europe did not interfere with the social round in England. In 1805, as Napoleon triumphed at Austerlitz, the Jockey Club offered the olive branch to the Prince of Wales. He had been racing again for five years but he forebore to return to Newmarket, contenting himself with building stables and a riding house attached to his Marine Pavilion at Brighton at a cost of £55,000.

Also in 1805 Lord Byron, who was a pupil and fervent admirer of John Jackson, played for Harrow in an impromptu cricket match against Eton and wrote a witty verse to gloss over the fact that Harrow were 'most confoundedly beat'. As Byron prepared to move on to Cambridge, so another enthusiast for the virility cult was moving on from Eton to Oxford. This was George Osbaldeston, the epitome of the sporting squire, the man for whom Chesterton's poem and Trevelyan's epigram might have been written.

Brought up by a doting widowed mother at Bath, Osbaldeston was heir to a Yorkshire estate, and from his earliest days was able to indulge his lifelong passion for betting and sport. His stay at Eton was brief and stormy, and his interests were not scholastic. 'I could beat any boy at single-handed cricket, or any boy of my age at fisticuffs', he afterwards claimed.[11] Other fashionable Etonian pursuits in which he indulged were shooting, fishing, and making squibs (once setting a servant girl's petticoats alight). At home he had received riding lessons from Dash, the most celebrated teacher of the day, and at school he hired horses in Windsor, or a gig to go driving; he also went to Ascot races.

Wellington denied ever saying that the battle of Waterloo was won on the playing fields of Eton. (He himself had played no games there: he played the violin and the nearest he came to an athletic encounter was a fight with a boy who objected to Wellington throwing stones at him while he was swimming in the river.) Indeed, in any meaningful sense there were no playing fields: Osbaldeston's miscellaneous collection of recreations was quite different from the team games that were gradually introduced in later decades – by the boys, not the masters – at Tom Brown's Rugby and elsewhere as an extension of the fagging system.

The nearest approach to organized sport at Eton of this period was rowing. In 1793 a young pupil wrote home describing 'a rowing and punting match set on foot by Her Majesty' on the King's birthday: 'there went up six boats, all with flags to them. . . . All the boys that pulled in them had caps with feathers.' In 1796 he wrote, 'There are four eight-oared boats and two sixes this year. . . . Tomorrow I shall put up in the third eight for Warren.'[12]

Rowing was one of Osbaldeston's accomplishments: he was to win many a bet in adult life. Of Eton he wrote, 'I belonged to the first rowing boat of the school and our crew could beat any other from Windsor Bridge up to Smiley Hall.' It was a rugged business, as befitted the virility cult of which it was part. 'Bumping one another's boats was the fashion then,' Osbaldestone recalled, 'but now it is called fouling.'

He left Eton at the authorities' request in 1803, aged 16, and spent two years at a crammer's near Brighton. He found time to acquire his first pack of hounds and to engage in horse races and short steeplechases on the Downs. He also, by his own account, inaugurated hurdling. 'One day while on our way home after hunting we came across some hurdles put up to pen sheep and proceeded to jump them. . . . The shepherd became very abusive.' As similar stories were told of the Prince of Wales it was small wonder that the Sussex shepherds were testy.

At Oxford Osbaldeston kept two hunters, as well as indulging in a bit of poaching. Old Etonians had also introduced rowing, though not at a serious competitive level. At this time it was mainly a social pursuit, as an undergraduate of 1805 recalled: 'Men went down indeed to Nuneham for occasional parties in six-oared boats (eights being then unknown) but these boats (such as would now be regarded as tubs) belonged to the local people.'[13] Osbaldeston, up at the same time, was more forthright: 'I kept up my rowing. Occasionally after going on the river a party of us would dine at the Star and Garter (a first-class hotel) and we generally got drunk; under these conditions we sometimes got into fights and rows with the townsmen when returning to our colleges.' Another favourite port of call was the King's Arms at Sandford. Osbaldeston was not popular with the authorities. Apart from his lawlessness he annoyed them by his pugnacity: the Head of his College once told him, 'I can't bear a Yorkshireman because he always offers to back his opinion by a bet.'

The year 1805 was a good one for the prize-ring. Henry Pearce, 'the Game Chicken', another Bristolian, beat Belcher for the championship and later that year beat an up-and-coming butcher's son – again from Bristol – John Gully, who was brought out of debtors' prison for the occasion. He was backed by Harry Mellish, the Prince of Wales's godson and a notorious gambler who was heir to large Yorkshire estates. Mellish had absconded from Eton in 1797 to join the 18th Light Dragoons, but coming into his vast fortune in 1801 was given permanent leave. It took him five years to spend it before a disastrous St Leger obliged him to join Wellington in Iberia. Meanwhile he and the Duke of Clarence secured Gully's release and 'a brother of Sam Barnard, the jockey, held a plate at the "One Tun" in Jermyn Street to collect money for a pair of breeches and silk stockings to fight in'.[14]

Thomas Creevey, a social and political aspirant, later recorded his recollections of the occasion in his diary. He was at a dinner in Brighton on the day of the fight:

> The Duke of Clarence was present at it, and as the battle, from the interference of Magistrates, was fought at a greater distance from Brighton than was intended, the Duke was very late. I mention the case on account of the change that has since taken place as to these parties. Gully was then a professional prize-fighter from the ranks and fighting for money. Since that time the Duke of Clarence has become the Sovereign of the country, and Gully has become one of its representatives in Parliament.[15]

Nelson's great victory at Trafalgar in 1805 was at the cost of his life, and the nation mourned. Pitt's own death in 1806, hastened by Napoleon's continued success, occasioned less grief. Indeed the United Irishmen and other dissidents found it cause for celebration when the new administration, chiefly Whig, let many of their supporters out of gaol. The government, optimistically called 'the Ministry of all the Talents', was successful neither in waging war nor making peace during its year in office and is chiefly memorable for abolishing the slave trade. 'I suppose you have heard', wrote the poet Thomas Moore to a friend, 'that during the Talents' administration Windham received an express from Lord Grey which made a sensation in every town it passed through, but which turned out to be the "announce" of a battle between Gully and Gregson, sent by the Foreign Secretary to the War Secretary.'

That was the year of a famous cricket match, a double wicket affair between Lord Frederick Beauclerk and George Osbaldeston, now a tearaway fast bowler at Oxford. Beauclerk was partnered by Howard, a professional, and Osbaldeston by an up-and-coming, but not always honest, professional called William Lambert. The match was for 50 guineas, 'pay or play'. Osbaldeston was ill but Beauclerk insisted on the letter of the law. Osbaldeston staggered to the wicket unable to bat but seeking to secure a substitute fielder. The man of God refused him even this, so Lambert had to play on his own. Indeed, he managed to win, having put Beauclerk into a temper by deliberately bowling wide at him.

Cricket's reputation was not enhanced in the following year when a gentleman named John Willes introduced round-arm bowling which was considered illegal, especially by batsmen. As Pycroft put it: 'Mr Willes and his bowling were frequently banned in making a match, and he played sometimes amidst great uproar and confusion. Still he would persevere until the "ring" closed in on the players, the stumps were lawlessly pulled up and all came to a standstill.'

The range of gentlemanly diversions was extended at this darkest hour of the war by the emergence of elite coaching clubs. The Benson, led by Sir Henry Payton, was founded in 1807. The road from Oxford to Benson, where the inn gave the club its name and headquarters, would be lined on meeting-days with the most elegant private coaches modelled on the mail coaches, and driven by young gentlemen expensively dressed in elaborate imitation of the coachman's garb. The Whip, founded the following year by Mr Charles Buxton, was a London club which adorned the streets in the vicinity of Cavendish Square. The *Sporting Magazine* described a typical meeting in 1810: 'The windows of the Nobility displayed a brilliant assemblage of beautiful females. Mr Charles Buxton's front drawing room, in particular, was crowded with rank and fashion. . . . Every dashing pupil of the new school appeared anxious to be seen – tandems, barouches, landaus, and, in short, every tasteful vehicle in London was driven to the scene.'

After the concourse came the drive. The delight of the upper classes in playing at what their inferiors did for a living often meant that liveried servants sat snugly inside a coach while the young master braved the wind and weather outside. Some Corinthians even went so far as actually delivering the mails. Osbaldeston, a Whip Club member, claimed to have a good deal of experience on the London to Brighton run. The coaching clubs were not supposed to race each other but of course they did. This kind of thing did not please the magistrates; but the coaching club members *were* the magistrates or their friends and relations.

Osbaldeston attained his majority in 1808. The occasion was marked by a set of homespun verses.

> The music sweetly played, and the bells did merrily ring
> Full bumpers overflowed to great George our King
> And to this noble square and the lady they thought best on

Unto this great lady Mrs Osbaldeston.

Naturally there were sports, no doubt supervised by the young Squire himself:

Smocks waistcoats and antkerchevs was run for on the flat
George Knaggs from Sherburn came and took away the hat.

And there was another important anti-revolutionary influence, free beer for the tenants and servants. As Osbaldeston recalled, 'Our guests partook so copiously of the ale provided that more than half of them were beastly drunk.'

Osbaldeston had 10,000 acres, stocked with grouse, pheasant, hares and rabbits, and a small deer park. He was an expert shot, but his chief passion was hunting. 'It was about this period,' he wrote, 'that I built Kennels on Beckford's plan . . . and bought my first pack. This consisted of the long-eared blue-mottled Southern Hounds which I brought out of Sussex. I hunted them for a time, but became disgusted with their independence and their way of hanging on the line.' They were good at the intricate hare-hunting work but were much too slow for the young Squire, who acquired a pack of dwarf fox-hounds and was soon racing them against a hunting parson's pack.

Osbaldeston was to spend little time on his squirely duties, and he soon squandered his inheritance on sport and betting on sport. But he epitomized Squiredom. The story goes that in later years a man pointed him out to a friend, saying: 'Look – there's the Squire'. 'Squire of where?' asked the friend. 'Squire of where? Why, he's the Squire of England.'

IV

Fortunately for Britain Napoleon's imperial urges were beginning to prove counter-productive. His invasion of Portugal and Spain aroused such patriotic reaction there that British hearts were stirred and uplifted. The call to battle, led by Canning, the Foreign Minister, was echoed by such diverse people as William Cobbett, a middle-aged radical, the intellectual Coleridge and the highly popular romantic poet Walter Scott. It was a timely revival. On 24 May 1808 thousands of starving Lancashire weavers, ruined by the blockade that was keeping out American cotton, invaded Manchester and were only dispersed by the intervention of the 4th Dragoons and the local yeomanry.

By a series of accidents Wellington (then Sir Arthur Wellesley) found himself in Portugal and eventually heading the expeditionary force. His men were, he reckoned, 'a rabble'. Poaching was the least of their villainies: they were desperate with hunger. Life was more agreeable for the officers, and Wellington encouraged them to keep fit in hunting, shooting, coursing and horse races. He drew the line at letting them go home for the winter to look after their estates, but he was ready to dispense with the wilder elements who had a thirst for 'deep play'. Harry Mellish, who had done well in actual fighting, was so addicted that he had to be sent home.

In England the talk was of the new boxing champion, Tom Cribb. Cribb, variously a bell-hanger, coal-porter and sailor, in an early victory over Belcher had attracted the notice of a Scottish gentleman farmer, fitness fanatic and entrepreneur called Robert Barclay Allardice, generally known as Captain Barclay. Barclay, who started a fashion for freakish feats of athletic endurance – at Newmarket in 1809 he walked a thousand miles in a thousand (non-consecutive) hours – also introduced revolutionary training methods into the boozy world of prize-fighting. When he sponsored Cribb for a return match with Belcher in 1809, he made sure his man was in good condition, and as Gully had retired Cribb claimed the championship.

Gully had his admirers, but he had not been a popular champion, partly because he was unreliable. Also his appearance in the lower circle at Drury Lane in 1807, a blatant attempt to rise above his station, drew indignant comment from the newspapers. On retirement he took up the lucrative but socially suspect trades of publican, coal-dealer, and, chiefly, bookmaker. It was Gully, according to the well-connected but by no means unblemished racing manager, Charles Greville, who began the 'system of corruption of trainers, jockeys and boys, which put the secrets of Newmarket at his disposal and in a few years made him rich'.[16]

Corruption thrived in the extravagant atmosphere of the Turf. The Derby never had fewer than 30 entries during the war, and many new races, later to become famous, date from the period – the Jockey Club Stakes and the Doncaster Cup, 1801; the Ascot Gold Cup, 1807; The Two Thousand Guineas, 1809; The Goodwood Cup, 1812; and the One Thousand Guineas, 1814. Both Gully and William Crockford, a former fishmonger, accrued great wealth through dubious practices, and the Turf had its godfathers, Jim and Joe Bland, strongly suspected of organising the outbreaks of poisoning that hit racing stables in the later war years.

In his early days as a bookmaker Gully investigated the possibilities of lining his pockets through cricket. He lacked the patience to learn the ins-and-outs of the game, but his interest indicated that cricket had become big business: important matches drew crowds of 4,000–5,000 and gates could reach £2,000. Players' wages were 'six guineas to win and four to lose or five and three if they lived in town.'[17] Matches were got up through subscription lists posted in the gentlemen's clubs, and Tom Lord did very well out of his investment. His profits were threatened in 1809 when he was asked to pay a bigger rent for the ground as a condition of renewal of the lease. London's expansion meant that land suitable for housing was at a premium. He declined to pay, and acquired another ground. The *Morning Post* announced: 'CRICKET GROUND . . . LORD begs to inform the Noblemen and Gentlemen that he has levelled and enclosed at the top of Lisson Grove, a short distance from his old Ground, which for size and beauty of situation cannot be excelled. . .'. The Marylebone Cricket Club were not convinced, and declined to move.

Urbanization was also a problem for the hunting fraternity. The Old Berkeley Hunt (OBH) found themselves in regular conflict with their neighbours. In 1808 a meeting of 'Noblemen, Gentlemen and Farmers' met under the Chairmanship of Lord Essex at the Abercorn Arms, Stanmore, to discuss 'the very serious evils resulting from the practice of hunting with the Berkeley hounds in the vicinity of the Metropolis. . .'. Personal issues clouded the conflict. The Master of the OBH was the Rev. William Capel, who had the cure of souls at Watford, and was Lord Essex's half-brother and his sworn enemy. In 1809 the OBH hunted a fox into the Essex estate at Cashiobury, and Capel and a follower jumped over locked gates, and smashed the top rail of a fence. Lord Essex sued them for trespass and the judgement in his favour took the traditionalists by surprise.[18] The hunting people took it for granted that they had the right to rampage over other people's property. Other suits followed, but not very many. Most people who owned land hunted themselves, and most tenants were either sympathetic or prudent.

V

The hypocrisy of the fashionable sporting set continued to attract critical comment. But there were more serious problems for the poor. There had been, and were still to be, years of boom as well as recession for industry; blockades and counter-blockades took their toll and even coal and iron had their recessions. Throughout it all the basic problem was the price of bread. Before the war wheat had been 43 shillings a quarter: by 1810 it was on average 103 shillings. The bad harvest of 1809 threatened the country with mass starvation, a condition averted mainly because the following year Napoleon yielded to pressure from his farmers to allow them to sell their surplus

1.   Cribb v Molyneux, 1811 (cartoon by Thomas Rowlandson, British Museum)

to Britain.

That winter, as Wellington lay entrenched behind the lines at Torres Vedras, the talk at home was of the match between Tom Cribb and the black American, Tom Molyneux, a freed slave. Molyneux encountered prejudice, not so much for his colour as for his nationality. The ex-colonials were being very troublesome politically and economically and boxing was John Bull's own sport. The affront was the greater when the American came within an ace of winning. Cribb had been indulging himself in London and was not in the best of condition, grossly overweight. Only the English weather saved the champion. It was December and bitterly cold. With Cribb in trouble his seconds began a dispute with the judges. The long delay gave Cribb time to recover and left poor Molyneux shivering. When the fight resumed he got cramp and had to quit.

Wellington's struggle in Iberia was beset by fears that political changes at home would betray them. With George III mad again, imagining himself hunting and hallooing with the hounds, a Regency was essential, but the general expectation was that, given power, the Prince of Wales would follow the Whig line and recall the expeditionary force. On the other hand, attempts to limit the Prince's authority would certainly lead to the dismissal of the Tory cabinet. The new Prime Minister, Spencer Perceval, took courage in both hands and surprised himself and everyone else by getting the Prince's support.

The Regent nevertheless spent the following summer at Brighton, cut off from his ministers. The *Sporting Magazine* took the opportunity to criticize him for neglect, not of the war but of the local cricket team: at one time the Brighton XI had been a match for 'all England and the Mary-le-bone Club' but 'as nothing flourished in this part of the world but in the renovating rays of the Heir Apparent, our cricketers had no sooner lost the fostering influence which could alone inspirit them into action . . . than supineness succeeded and they at last . . . dwindled and degenerated into their original insignificance'. A more serious sporting scandal that summer was a further outbreak of poisoning in racing stables, aimed at stopping certain highly fancied horses. The Jockey Club put up reward money of 500 guineas. They did not manage to connect the poisoning with the Blands, but a tout, Daniel Dawson, was convicted and duly hanged at Cambridge before ten thousand spectators.

Almost as many, including a great number of persons of quality, journeyed to Thistleton Gap in Leicestershire to see the return bout between Cribb and Molyneux. Cribb was again in the hands of

Captain Barclay who took no chances with his investment, taking him to Scotland to train. Cribb crushed his opponent. He picked up 400 guineas and Barclay, according to the *Dictionary of National Biography* made £10,000.

Better progress in the war in no way lessened government fears of Jacobin conspiracy at home. It was a time of violent industrial conflict, and the early fighters for workers' rights found themselves driven underground and infiltrated by government counter-spies. The Luddite riots that began in 1811 signalled the emergence of an organized alternative culture. 'In the heartlands of the Industrial Revolution,' as E.P. Thompson put it, 'new institutions, new attitudes, new community-patterns were emerging which were, consciously and unconsciously, designed to resist the intrusion of the magistrate, the employers, the parson or the spy.'[19] The Combination Acts had finally forged the link between trade unionism and sedition. Luddism was the reaction of textile workers, especially the skilled, to the prospect of being replaced by machines. Rioting, the old method of redressing wrongs, was no longer effective, and in Nottinghamshire, marauders under a mysterious leader, Ned Ludd, based in Sherwood Forest, began to make organized, secret raids to smash the dreaded machines. The troops were powerless against an invisible enemy and it was left to undercover agents to provide the authorities with information. Once they had it, the magistrates hit back with sentences of death or transportation.

The severity of the Game Laws was a persistent source of resentment and trouble. Even supporters of the establishment thought it unfortunate that tenants were not allowed to shoot or hunt on their own farms or to protect the game from poachers. 'The vagabonds', argued the lawyer Joseph Chittey, 'are rather encouraged than prevented by the present laws. For if the Farmers who live upon the lands . . . are not allowed to sport themselves or to be game-keepers of their Farms . . . they will either destroy the Game out of resentment or allow the poachers to do it for them. . .'.[20] But the oppressive system continued.

Another bad harvest in 1811 increased social tension. A drastic drop in exports, Irish troubles, high taxation and political faction added to Spencer Perceval's unpopularity. He stuck it out bravely, and things were improving when, in 1812, he was assassinated. His death was greeted with delight by mobs in London and the north and it was felt unwise to give him a public funeral. Perceval's successor, Lord Liverpool, the great committee man, soon found himself fighting two wars as worsening relations with America spilled over into conflict, but the new administration was staunch in its support of Wellington.

As he hatched his plans to drive the French from Spain, Wellington's officers – 'sport mad' according to one observer – enjoyed beagling, fox-hunting and even an occasional wolf-hunt. Wellington himself spent much time running foxes to earth, but apparently mainly for the exercise. Throughout the winter he reputedly took only one fox when his hounds 'mobbed' it. Still, he evidently hoped for better things. In March 1813 he was told that Lord March had applied to Lord Easton who had found him a suitable pack in Buckinghamshire.[21]

Protracted negotiations were also successfully conducted with Rothschild for funds to continue his campaign, and things seemed much brighter at home and abroad. The good harvest of 1813 helped to discourage Luddism, and the price of corn tumbled. It was the end of easy profits for farmers, but they still had a little fat to live off.

The invasion of France began in the spring of 1814. The victory at Toulouse was costly but decisive. The Allies, led by General Blücher, entered Paris, Napoleon abdicated, the Bourbons were restored and Wellington became a Duke.

VI

In England the peace celebrations, although premature, were greatly enjoyed. The Derby attracted a record entry of 51, and Alexander, King of Prussia, Tsar Alexander I of Russia and General Blücher were there to see the aptly named 'Blucher' win. The enthusiasm was rapturous. Later, at Ascot, where Blücher attended the Gold Cup, even the Prince Regent was included in the cheers. Another celebratory event was an exhibition of pugilism at Lord Lowther's residence in Pall Mall. Tom Belcher, Tom Cribb, Bill Richmond and John Jackson showed their skills before the visiting dignitaries.

Boxing was now an acknowledged British art, and its devotees sought greater respectability. In 1814 the Pugilistic Club was set up. Its 120 founder members guaranteed an annual sum from which purses of 10–50 guineas were offered for public contests. Apart from ensuring patronage, the club's aim was to try to clean up the sport. Its articles declared 'the expressed determination of the members to expose all crosses and to prevent . . . transactions of this kind.' They introduced other innovations. A roped square was constructed for the bout between Bill Richmond and the Irish 'navigator' Davis in 1814, the first time such a 'ring' had been used. Bouncers had distinctive uniforms, dark blue ribbons in their hats, to avoid the confusion that otherwise arose when stewards tried to control the crowd. The members themselves had blue and buff uniform coats, with yellow Kerseymore waistcoats and buttons engraved with 'P.C.'. Sir Henry Smith took the chair at the first annual dinner when Lord Yarmouth was the main speaker. As an admiring account said of the Club: 'The high respectability which they confer by the patronage of their rank is of inestimable benefit . . . also by the substantial rewards to valour which their united funds enable them to bestow.'[22]

It was a good year for cricket, too. The MCC had never liked Lord's Lisson Grove ground, even though he had transferred the original turf there. Now he was obliged to seek another site because of the proposed route of the new Regent's Canal. He found one in St John's Wood, lifted his precious turf once more, and erected a tavern, a pavilion and various outbuildings. There were to be no more moves for the MCC, but a bizarre incident took place a few days before the opening. The *St. James's Chronicle* of 5–7 May announced 'A shocking accident occurred on Thursday at the New Lord's Cricket-Ground public house. The landlady of the house had occasion to use a small quantity of gunpowder, and whilst in the act of taking the same from the paper, containing a pound weight, a spark from the fire caught it and it went off with a great explosion'.

No alien spies, it seems were involved, and with Bonaparte in exile and the Americans, after the sacking of Washington by excessively zealous British troops, ready to make peace, all seemed ready for a return to normality. For some this meant unemployment, as war demand ceased; and with corn at low prices many small farmers went bankrupt, and a number of country banks were in difficulties. Wellington's duties as Ambassador to the Court of the Tuileries included hunting with the Royal family, an elaborate and stately pursuit in which the professional huntsman did the work. The Bourbons lived up to their reputation of forgetting nothing and learning nothing.

In March 1915 Napoleon escaped from Elba and marched on Paris without a shot fired in resistance. As the tension mounted during the Hundred Days, an army was assembled for Wellington only with difficulty. The Radicals at home were gleefully expectant of a Napoleonic victory, and in Brussels the Duke had to instil calm into his potential allies. On 12 June he took one of the Duke of Richmond's daughters to a cricket match at Enghien. Another source of relaxation was racing. Lord Uxbridge, Sir Hussey Vivian and the other cavalry commanders were prominent in meetings organized by Arthur Shakespeare, a young Dragoon officer, who three times acted as Clerk of the Course at meetings of Grammont. The Duchess of Richmond's ball on the eve of Waterloo was the climax of this insouciant approach.

Other insouciant Dubliners were less concerned with the fate of their distinguished fellow-citizen than with the prowess of the prize-fighter Dan Donnelly. Donnelly, an Irishman nicknamed 'Sir Dan' after the Prince Regent had clapped him enthusiastically on the shoulder, would perhaps have been the greatest fighter ever had he been able to abide by Captain Barclay's principles. He had particular trouble with two of them: abstinence from ardent spirits and sexual intercourse. Dan liked to take a bottle to bed with him to prevent his lying awake – and he did not mind whose bed it was. Venereal disease did not help his fitness.

Nevertheless he was a fighter to be reckoned with, and a great favourite with the crowds. In 1814 he had lost on a disputed decision to the Englishman George Cooper and all Dublin was agog to see the re-match in 1815. The thousands who gathered in a nearby natural amphitheatre saw an epic match. It was commemorated in triumphant if inelegant verses:

> You sons of proud Britannia your boasting now recall
> Since Cooper by Donnelly has met his sad downfall,
> Out of eleven rounds he got nine knockdowns, beside a broke
>   jawbone
> Says Miss Kelly, 'Shake hands, brave Donnelly, the victory is all
>   won'.[23]

His victory was all too complete: in December 1815 the *Sporting Magazine* reported that 'Cooper the pugilist who was vanquished by Donnelly . . . died a few weeks since, supposedly from the tremendous blows he received on that occasion'. Reduced by the rackety life he had led, Donnelly did not survive him long. The final assault on his frame was drinking a vast quantity of cold porter after playing fives. *Blackwood's Magazine* commemorated his passing with a mock-heroic 'Luctus on the death of Sir Daniel Donnelly' with poems in Greek, Latin, Hebrew and English in imitation of the great poets.

Manly sports, indeed, enjoyed much support in the widening literary circles of the age. Hazlitt, though critical of the animal-baiting side of John Bull's nature, was enthusiastic about boxing and fives-playing. Borrow was to grow lyrical about the bruisers of England. George Cruikshank, a devotee of rowing and boxing, illustrated the description of the Cribb-Molyneux return for his friend

Pierce Egan, the chronicler of the 'Fancy' whose *Boxiana* began publication in 1812.

By the time Britain fought its next major war a century later, the virility cult had been taken up and transmuted by the public schools. Team games were invested with moral values and the capacity to develop qualities of leadership, and the code required that in wartime these qualities should be put at the service of the nation. The way the various sports responded to the call became a major issue. The rugger men and cricketers gave up organized competition for the duration, and professional soccer and the Northern Union yielded to the class-based criticism of the moralists. But after the 1917 season *Wisden* complained of the illogicality of ending cricket 'with public boxing carried on to an extent not heard of before'. Jimmy Wilde became World Champion in 1916. Racing also withstood criticism and carried on throughout the war. Among the defenders of the Jockey Club was the fifth Earl of Rosebery who pointed out that the Ascot Gold Cup had been run eight days before Wellington met Napoleon at Quatre Bras.

NOTES

1. John Wardroper, *Kings, Lords and Wicked Libellers* (London, 1973).
2. G.M. Trevelyan, *English Social History* (London, 1944).
3. Ibid.
4. 'Derby Mercury', quoted in F.P. Magoun, *A History of British Football from the Beginnings to 1871* (Cologne, 1938).
5. 'The Perambulations of Barney the Irishman', quoted in Percy M. Young, *A History of British Football* (London, 1968).
6. See John Arlott, Introduction to John Nyren, *The Young Cricketer's Tutor* (London, 1833; new edition 1974).
7. *c*. 1792.
8. Quoted in Roger Mortimer, *The Derby* (London, 1973).
9. Joseph Strutt, *Glig-gamena angel deod; or, The Sports and Pastimes of the People of England* (London, 1801).
10. The Rev. James Pycroft, *The Cricket Field* (London, 1851).
11. E. Cuming (ed.), *Squire Osbaldeston: his autobiography* (London, 1926).
12. Quoted in Hylton Cleaver, *A History of Rowing* (London, 1957).
13. G.V. Cox, *Recollections of Oxford* (London, 1868).
14. E.H. Budd in C.A. Wheeler (ed.) *Sportascrapiana* (London, 1867).
15. John Gore (ed.), *The Creevey Papers* (London, 1963).
16. Henry Reeve (ed.), *The Greville Diaries* (London, 1875).
17. William Ward, quoted in Pycroft, op. cit.
18. E.W. Bovil, *English Country Life 1780–1830* (Oxford, 1962).
19. E.P. Thompson, *The Making of the English Working Class* (London, 1963).
20. Joseph Chitty, *Essays on the Game Laws* (London, 1969).
21. Elizabeth Longford, *Wellington: the Years of the Sword* (London, 1969).
22. Quoted in John Ford, *Prizefighting* (Newton Abbot, 1970).
23. Quoted in S.J. Watson, *Between the Flags* (Dublin, 1969).

3

Social Stratification and Participation in Sport in Mid-Victorian England with Particular Reference to Lancaster, 1840–70

M.A. Speak

I

During the last decade, social historians[1] have described and analysed an increasing variety of organized sporting and recreational activities in Victorian England and have paid particular attention to working-class and middle-class participation. Baker[2] has argued that to make sense of recent British sport historiography, however, analysts need to adopt Lytton Strachey's *Eminent Victorians* strategy:

> To row out over that great ocean of material, and lower down into it, here and there, a little bucket, which will bring up to the light of day some characteristic specimen, from those far depths, to be examined with a careful scrutiny.

There is some doubt though as to whether such a simple sampling technique is adequate, for the dearth of local and regional studies could well distort the true picture. Small sampling investigations must have very clear aims and objectives, and are subject to experimental error to say nothing of interpretive distortion. What is needed is not a bucket, but a drag-net over the whole seabed of Victorian sport and recreation before conclusions can be drawn which could in any sense have the value of the qualitative analysis likely to emanate from the regional surveys undertaken by Arnd Krüger in Saxony, West Germany[3]. Attempts have been made to depict a national picture, but Davis[4], having considered the work of Rees[5], Meller[6] and Molyneux[7] has claimed that '. . . such is the diversity of their findings that only when the majority of towns has been specifically researched can one hope to recognise a national picture.'[8]

Gray[9] has argued that analysis within a local framework can lead to a necessary narrowing and sharpening of focus, but in another sense can broaden the field of vision by allowing a clearer picture of the network of formal and informal institutions that shaped the Victorian social world. Walton and Walvin[10] have argued that the study of chosen themes in local settings can lead to the writing of a new kind of national history, and certainly any meaningful qualitative second-order comparative study must rely on carefully researched patterns of regional and local study. Indeed, there would clearly be support from coal-face historians for the view that a classification system should be created to allow comparisons between the very different forms of sporting organization and participation in industrial cities, county towns, rural communities and even seaside resorts and spa towns.

Seminal work has already been undertaken on participation in individual sports in Victorian Britain. Mason[11] for association football, Cunningham[12] for rifle shooting, Mandle[13] for cricket, Mawdsley[14] for athletics and Gray[15] on recreation in Edinburgh among others have all made original contributions, although there would appear to be a dearth of studies on participation in sport and social class across broader social boundaries. The work of Metcalfe[16] and Bouchier[17] in Canada does not appear to have a counterpart in Britain.

Participation in sport in developed societies is the direct or indirect result of a whole series of influencing factors. Socio-economic and marital status and mobility, age, sex, religion, education and the availability of leisure time and leisure facilities all affect both the quantity and quality of participation in sport and recreational activity. The investigations of the Sports Council into participation in sport in Britain in the 1970s and 1980s will allow future historians a very clear insight into the role played by sport and recreation within the social life of late twentieth century Britain. Much less clear, however, are the patterns of participation in mid-Victorian England. The growth of participant sport has been charted by several sport historians, and Rees[18] for Liverpool, Molyneux[19] for Birmingham, Davis[20] for Worcester and Elsworth[21] for Blackpool and Southport have clearly indicated that explosive growth occurred between 1870 and 1895, the result largely of increased leisure time, particularly the half-day Saturday holiday, together with improved socio-economic circumstances.[22] Less clear at present are the origins and organizational structures of sport at local levels before 1875 and precise patterns of participation.

In this contribution to the history of Victorian sport and to the eventual achievement of a national synthesis, an attempt is made to consider participation patterns in several sports in Lancaster and district against a socio-economic backcloth in the period 1840–70 which appears to bridge the decline of barbaric, often spectator-oriented recreations, and the advent of what middle-class reformers considered more civilized and civilizing participant-oriented sports. At the outset some consideration is given to social structure within the City of Lancaster and, wherever possible, for the sake of realistic comparison, the years close to 1860 have been used to extract participation details. A detailed scrutiny of several sports which serve the social needs of different citizens attempts to confirm or deny the views of social historians on whether or not sport served to reinforce social fusion or social fission.

II

Owen[23] has argued that social stratification refers to the ranking of people in terms of superiority and inferiority into a relatively stable hierarchy, according to actual or assumed possession of certain characteristics. This ranking has consequences for the life experiences of people thus ranked, and as a result of cultural transmission these rank orders are likely to persist over long periods of time. Storch[24] expresses the view that there was deep concern over the absence of a commonly affirmed array of values linking these different ranks in Victorian society, and that common leisure activities in the large urban areas often foundered on the rock of class conflict, with the result that each class, and even sub-class, was forced into its own distinctive leisure and recreational activities. Others have expressed the view that middle-class reformers of the period believed that social harmony was possible both in and through the world of sport.

Future research will no doubt provide the flesh for the skeleton of this dispute through a series of local and regional studies. This particular study involves Lancaster, a county town with 14,324 inhabitants according to the 1861 census, at the heart of a stable agricultural community. It suffered economic stagnation in mid-century and ceded trade institutions and entrepreneurial leadership to larger cities in the southern part of the county, in particular to Liverpool and Manchester. It failed to establish the powerful industrial base achieved by many south Lancashire towns, and its socio-economic structure comprised a smaller and more acquiescent industrial population, a large retail and handicraft sector and a larger-than-average professional group, based on medical and legal institutions, which was to have particular importance in the context of this study.

Researchers have clearly expressed the difficulties facing attempts to accurately classify large urban populations. Heiton,[25] writing in the mid-nineteenth century, characterized Edinburgh as a society of castes, with aristocrats, top professionals, merchants and shopkeepers all seeking to defend entrenched positions, and Gray[26] has noted the massive cultural differences separating labour aristocrats from lower middle-class strata. There are similar divisions within the landed gentry, the aristocracy, gentry and higher yeomanry, all enjoying (or hating) well defined social status. Given these particular problems, established work has been used to provide a framework for the study. Anderson[27] and Armstrong[28] have each gone to considerable lengths to break down earlier populations into more acceptable modern categories, the former from a sociological, the latter from a demographic, standpoint. Gooderson[29] has drawn on the work of both to illustrate the social and occupational structure of Lancaster in 1861, but this study concentrates on Armstrong's classification (Armstrong's Social Class) since its six-sector breakdown is rather simpler for our purposes.

TABLE 3.1

THE DISTRIBUTION OF LANCASTER HEADS OF HOUSEHOLD BY
ARMSTRONG'S SOCIAL CLASS (10% SAMPLE), 1861

		ASC		
1 and 2	3	4 and 5	6	Total
18.9 (56)	52.0 (154)	22.3 (66)	6.8 (20)	100 (296)

Anderson categorizes social class under six main headings: ASC 1: top professionals, entrepreneurs and employers of a work-force of more than 25; ASC 2: lower professionals and shopkeepers with one or more domestic servant; ASC 3: skilled craftsmen and small shopkeepers (a massive group); ASC 4: semi-skilled; ASC 5: unskilled; ASC 6: unclassifiable.

Again at the outset, some clear identification is perhaps needed of the range of sports practised in Lancaster during this period, and detailed local newspaper surveys reveal the following reported in varying degrees of intensity.

Sports regularly reported included those which attracted both participants and spectators, occasionally in large numbers. The list includes the following (the dates indicate the earliest newspaper report during the period): archery (1840), cricket (1842), cycling (1870), old English or popular sports (1840), wrestling (1840), football (1868), rifle shooting (1840) and swimming (1865). Other sports of a calmer nature which were not spectator-oriented were angling (1846), bowls (1840) and skating (1841), and there were the traditional animal-based sports of coursing (1840), hunting (1840) and horse racing (1840). Those sports occasionally reported including cock-fighting and prize-fighting, which were basically spectator-oriented, and the popular and traditional sports of quoits, knurr and spell and pitch and toss.

The sporting calendar was regular, conditioned by seasonal influences, and the yearly cycle of sport consisted of coursing and hunting in January, February, March, October, November and December, with rowing, bowls, horse racing, cricket, archery and athletics falling within the period from March to October, and finally shooting, popular sports, wrestling and pedestrianism appearing throughout the calendar year. Although seasonally regular, there was by no means massive local participant activity. This would not emerge until the 1880s and 1890s, when improved socio-economic conditions would provide the wherewithal to accommodate regular sporting activity and competition, often in the form of leagues, for a broader cross-section of the community.

III

Despite the loss of the assizes in 1835, economic stagnation and declining trading wealth, Lancaster retained the atmosphere of a county town. A strong industrial elite cemented the link between the city's wealth and the county gentry, and the emergence of distinguished professional figures like Edmund Sharpe, architect, Thomas de Vitre and Thomas Howitt, surgeons, was to have a supreme influence on the nature of, and participation in, sporting activities of the district. In the 1840s a strong impetus was given to the local social sporting calendar. Attempts by the local middle-class community to stimulate 'rational' forms of recreation and the intellectual vigour of the community by establishing the Athenaeum, reviving the theatre and setting up soirées and concerts went hand in hand with the introduction of a series of sporting events which contributed to a regeneration of local sporting activity and interest. The annual, socially prestigious, gala week in archery was supplemented by the establishment of clubs for cricket (1841) and rowing (1845); the racing calendar was reinstituted in 1841 by the Tories after threats of closure from a Liberal council in 1840, the regatta was revived in 1842, the Lancaster Hunt reconstituted in 1845 and the formation of the 10th Rifle Volunteers in 1859 restored the military associations, which had been shattered by the removal of the Yeomanry exercises, and paved the way for regular rifle shooting contests.

These events, added to existing activities such as hare-coursing and game-shooting, gave the upper leisured and professional middle classes a complete annual cycle of sporting and social events. Veblen[30] has argued that not only did the wealthy not work in order to maintain their honour but they also had to be seen to be idle. This consumption of time in a non-productive sense was characterized by a variety of cultural appendages including sport and physical recreation. Delves[31] has noted that local societies often lacked the flexibility to accommodate the working classes, since events had starting times and dates impossible for the working man, subscriptions were required in advance and there were elaborate membership procedures and specific requirements in terms of dress and decorum.

Investigations into the membership lists of local sports organizations confirm these views and reveal a not unexpected pattern of participation, given the exclusive, rather than inclusive, nature of organized sport. Activities researched in detail include archery, rowing and cricket. These attracted limited membership in Lancaster, although future studies in other regions may reveal completely different patterns of membership and participation.

The John O'Gaunt Archery Society's annual gala and field days were important events in the county's and city's social calendar. The season began in May with the election of a prestigious patron or patroness, the field and gala days in July and September were widely reported, more for their social brilliance than for any sporting prowess exhibited, and each occasion was followed by a dinner and ball, again extensively reported in the local press. Membership of the society was restricted, and scrutiny of membership lists (1840–70) reveals that identifiable members were gentlemen (13), surgeons and physicians (3), attorneys and solicitors (4), sheriffs (2), mill owners and senior managers (3), members of Parliament (1), merchants (2), rectors and clergy (2). Seen in tabulated form, the social exclusiveness is obvious.[32]

Exclusiveness was maintained by specific rules and regulations. The society was to consist of not more than 30 male and 20 female members; 'gentlemen desirous of being members' had to be proposed and seconded in writing and three votes against were sufficient to bar prospective members. The annual subscription was set at three guineas with an additional two pounds entry fee; badges of the society had to be worn on penalty of a ten-shilling fine, and field days were held on the first Thursday in June, July and August each year.[33] All of these clauses sought to preserve an élite society, and ensured that participation by the working classes was unthinkable, on both social and economic grounds.

Rowing was also a sport which was part of the social calendar, particularly after the revival of the regatta in 1845. At local level in mid-Victorian England, it consisted either of highly organized middle-class club events, which after 1845 increasingly took the form of inter-city club matches and annual regattas, or alternatively working-class rowing events linked with Whitsuntide and other popular holiday sports occasions with frequent wager races in the evenings.

Opportunities to participate in archery have been seen to be unavailable to those outside a certain class, but the sport of rowing, not an élite activity, allows us to examine the social schisms which often spilled over into the world of sport. In common with archery, restrictions on membership were applied to ensure exclusiveness. A blackball system operated to restrict membership to acceptable applicants who had to be proposed, seconded and accepted on a vote. There is no information available in the minute books on how detailed the information sought on applicants was, or what guidelines governed blackballing, but Bouchier[34] has revealed that in Ingersoll, the names, place of residence and occupations of the nominator and nominee were announced to the general membership before a blackball vote. The rules of the Lancaster club were framed by the committee at a meeting on 5 November 1842, and in common with those of the archery club included the payment of subscriptions, the wearing of a uniform, and a clause which required that 'Any member appearing without the uniform of the club at a match, or without his jersey for practice, shall be fined at the option of his coxswain'.[35] The committee at the inauguration of the club consisted entirely of ASC 1 figures, some of whom were also members of the committee of the cricket club and members of the archery society. By 1861, the committee consisted of gentlemen (2), Justice of the Peace (1), architect (1), schoolmaster (1), silk spinning manufacturers (3), solicitor (1), dispensing chemist (1), corn inspector (1) and wine merchant and general ironmonger (1); using Armstrong's classification we again are able to perceive a clearly identifiable pattern of membership.[36]

In the world of rowing nationally, exclusiveness was further maintained by carefully segregating competitors in local club regattas on a socio-economic or membership basis. In many club regattas, categories might be (a) members of the club (already limited on a socio-economic basis); (b) open to the world; (c) local residents; (d) amateur members of the club; (e) gentlemen amateurs and members of the club; (f) youths under 18 years of age; (g) women; (h) mechanics;[37] (i) fishermen.

In 1846 at Lancaster there was a dispute following a Manchester crew's victory in the Borough Cup. The essence of the dispute was that victory was challenged and the cup withheld, on the grounds that two members of the Manchester crew were not acceptable as entrants. Three of the crew, Martindale and Lambert, who kept butchers' stalls in Manchester, and Sumner, who was a letter-carrier, were acceptable. Peace, however, a fourth member of the crew, was a working cabinet-maker, and Lowe, a working bricklayer, and it was pointed out that 'they were not known as men of any property in Manchester'. *Bell's Life* was consulted and came up, with some obvious difficulty, with a ruling:

> Our opinion decidedly is that if Peace and Lowe are both master tradesmen as Martindale and Lambert appear to be then they are entitled to the Cup. If either of them are journeymen or mechanics, the prize must go to the second boat. We recommend the managers of this regatta to avoid the difficulty in future by adding to the conditions of the race that the gentlemen amateurs should be members of some regularly constituted amateur rowing club, as at the Thames and Henley Regattas.[38]

Historians have argued that attempts at class integration constituted a dominant ideology of leisure in mid-Victorian England. It would appear that this ideology was doomed to failure but this example would appear to show that whatever the ideology, in practice considerable social divisions existed not only within local society, but also within local and regional sport.[39] Two further examples serve to indicate the jealous nature with which social privileges were guarded.

First, several regattas at Henley, Chester, Lancaster and elsewhere were held in high esteem in the social calendar, but it is obvious that other regattas were frowned upon as being commercially oriented. The Minutes of the Lancaster club commented on the Manchester and Salford regatta, which according to the scribe was:

> . . . as usual a blackguard, disreputable affair, got up by a set of publicans and Pomona Gardens[40] proprietors for their own particular benefit. . . . The best Chester crew [socially acceptable to the Lancaster Club and highly regarded by them] entered for the Grand Challenge Cup and was beaten by 200 yds at least by the Manchester boats.[41]

The matter was of such concern that on 2 June 1853, the Secretary of the Lancaster club was requested to write to Henry Lloyd of Chester to ascertain the line of distinction between a 'gentleman amateur' and an 'amateur'. No further information was available in the minute books, however, although in 1866 the Amateur Athletic Association defined an amateur as a person who had *either*: never competed (1) in open competion; (2) for public money; (3) for admission money; (4) with professionals for a prize, public money or admission money; *or* (5) never taught or assisted in the pursuit of athletics exercises as a means of livelihood; *or* (6) was not a mechanic, artisan or labourer.

Second, local regattas connected with wagers, or limited to lower-middle or working-class competitors were increasingly frowned on by the club élite. Local working men held frequent races off the quayside close to the city centre, usually in the evenings when work was finished. The minutes of the Lancaster rowing club indicate the disparaging attitude towards these events by the middle classes. 'The Quayside people held their *mudlark* regatta today. A committee of Management (save the mark!) organised the affair . . . at night the committee, the competitors and a few "navvies" got hilarious together, but at whose expense this deponent saith not.'[42] Lists of names of competitors simply reveal that none of them were householders and it can only be concluded that they were working-class.

Lancaster Cricket Club was founded in May 1841 by eight middle-class citizens, some of whom were members also of the archery society and rowing club and their committees. A set of rules was printed in August and included several, akin to those for archery and rowing, which served to restrict membership to a limited social circle. New members were admitted by ballot, the annual subscription was one guinea and a system of fines operated for late payment of subscription and even for leaving a match before its conclusion 'without the unanimous consent of the whole field'. The club was to meet for practice on Tuesdays, Thursdays and Saturdays which was delimiting, and field days were to be held on Tuesdays in every fortnight with fines for non-attendance. The club uniform consisted of white trousers, a striped blue and white Guernsey woollen shirt, and a straw hat, with fines for failure to wear it.[43]

Membership of the club was limited on a socio-economic basis, although analysis of playing membership is interesting. During the period in question, 1840–70, there were three stages in the club's development, clearly identified by Gilchrist[44] which reflected the challenge facing privileged participants. The first period (1841–58) consisted largely of participation by upper middle-class professional figures. The second saw the formation of a new, lower middle-class organization, and the third (1865–78) a reversion to a more privileged membership.

During the first period, 1848–1858, playing records reveal a preponderance of ASC I and ASC II figures.[45] Gilchrist[46] has recorded the nature of sport in this period, wherein matches were played any day in the week over two innings, wickets pitched at 9 or 10 o'clock in the morning, lunches accompanied by toasts and speeches, dinners held at the end of a match, and he has described them as having 'a largeness and generosity about them'. The high proportion of members in ASC I and II can be readily understood, given the social nature of sport at the time, the time-tabling of matches and likely costs involved. An advertisement in the *Lancaster Gazette*[47] called for applications for membership from 'gentlemen' to the secretary[48] and the two players in ASC IV were in fact butlers, presumably only playing to make up the numbers.

The first period ended in 1855 when club members were scattered. In the second period a new club was established. The club scorebook, available to Gilchrist in 1909 but no longer in existence, had an undated comment in it written by L.T. Baines, club secretary: 'A pseudo-Lancaster C.C. was started about 1860 which generally lost its matches, and whose performances have for obvious reasons, been excluded from the register. This club included none of the original members of L.C.C. and was in no sense representative of the cricket of Lancaster'.[49] There had been reports in the local press in 1850 of a new club being formed – the Tradesmen's Club – and it could well be that the new club was an offshoot of this. The 50 or so members of the club were described as being 'men of a younger and different generation' and whereas previously and subsequently, matches were against socially acceptable opposition, the new fixture card included a match against Kendal Mechanics.[50] The make-up of the club at this point has been analysed and reveals that there was in fact a massive decline in the numbers of ASC 1 and ASC 2 participants, the bulk of the membership belonging to ASC 3.[51]

The third period (1865–78) saw a revival of the old social institution after the collapse of the 'bastard' ASC 3 club. Fixtures against the officers of the Duke of Lancaster's Own Yeomen Cavalry and the 1st Duke's Own Militia, the Grammar School and Sedbergh School, along with the usual inter-town matches, reproduced the social nature of an earlier period. Once again a preponderance of members belonged to ASC I and ASC II[52] and if the batting order of the Yeoman Cavalry is a yardstick, it may well be that in addition to social exclusiveness operating to exclude the lower orders, there was a very clear pecking order within the club, even on the field of play.[53] During this period, however, several local village clubs were established, local district school and church clubs were beginning to add to the number of cricketers, and there is no attempt to suggest that cricket in the village or within educational and religious organizations was necessarily as exclusively controlled as within the city club. There is evidence of two regular and nine irregular clubs within Lancaster and eleven regular and two irregular clubs in the outlying districts between 1861 and 1870. It is merely possible to hypothesize that there was a growing democratization of cricket during the 1860s, a process to be radically accelerated during the 1880s and 1890s as already indicated, but that democracy did not always apply within the more formal organized sport institutions. The sports already analysed have revealed their origins, administrative processes and participation to be mainly upper middle-class. There is little doubt that individual protagonists had not only the social, financial and professional status to enjoy sporting leisure, but called upon their experiences at public schools and universities to establish an ethos and order which became a springboard for later generations. Several of the identifiable figures in the clubs were educated at Eton, Sedbergh, Lancaster Royal Grammar School and later at St John's, Cambridge and Trinity College, Dublin.

Bowling was a well established local sport, changing its nature according to the demands of local society. In the eighteenth century there was a bowling green on the banks of the River Lune and the castle had its own green within the walls for the use of prison warders. In the mid-Victorian period, the opening of Mr Bagot's commercial bowling green at Luneside in April or May annually, and its closure in September were as regular and traditional, albeit in a lower social key, as the gala day of the John O'Gaunt bowmen. Between 1840 and 1850 the Luneside bowling green was an almost unique facility in the district, offering in-house tournaments, often with wagers and prizes. Between 1850 and 1860 however there was a growing interest in matches against individuals, or teams, from neighbouring towns and the period 1860–70 saw a growth in the availability of local and district greens (eight) and representative matches, always with prizes and money at stake. At the end of a representative match, individual wager contests were the order of the day. Matches were held on weekdays during this period, and an analysis of the Lancaster team which represented the town against Garstang on 31 August 1861 reveals that participants were mainly self-employed tradesmen, shop-keepers or inn-keepers. Precise occupations discovered were inn-keepers (3), butcher, shoemaker, cabinet-maker, auctioneer and manure agent, revealing that the sport was particularly attractive to the commercial classes. Matches involved travel to other towns, no small expenditure on entry fees, sweep stakes, meals and presumably alcohol. Prizes of snuff boxes, watches and cash were mainly available.[54] An analysis of the Lancaster team which played Garstang in 1861 reveals that all the players were from ASC 2 and ASC 3, although it is recognized that the numbers involved are so small as to make serious conclusions tentative.[55] It is again evident, however, that the necessary availability of time and money to participants would of necessity place limitations on participation by the lower ASC groups.

IV

Both Weber and Veblen[56] have argued that social inequality in society can be observed by noting the life chances of individuals. Members of the higher social classes could monopolize luxury goods and leisure opportunities to secure a distinctive and prestigious life style. During the 1840s, a period of political and social unrest, not only were these privileges jealously guarded, as has already been indicated, but attempts were also made by the middle classes to delimit and defuse popular sports. Storch[57] has drawn attention to the apprehensions of the middle classes over the nature of working-class leisure and Vamplew[58] has indicated the changing nature of popular sport in society in its relationship with forms of industrialization, and in particular the ways in which employers supported a general middle-class policy of imposing a morality on both idle rich and idle poor. Since respectability was the aim of society, campaigns were mounted against drink, idleness, dirt, popular customs and sport. Work on the one hand was fixed, immutable and directly supervised, but leisure and sport on the contrary posed a challenge to middle-class authority and status. Indeed the riot at Lancaster Race course[59] on 23 July 1840 provides an illustration of the threat to social order and control of crowds gathering in large numbers at sports events. During a period of social unrest, when strong working-class consciousness was enhanced by a plethora of movements for social reform, whose very existence and militancy suggested a general discontent on the part of working people, it is easy to appreciate the fears aroused by sports gatherings on the part of an equally class conscious but conservative middle-class.

A whole series of controls operated, either in direct form, such as bye-laws and the vigilance of the new police authorities, or indirectly through the promotion of rational forms of recreation, membership procedures and the influence of religious authorities. The middle classes saw it as their responsibility to fill the vacuum created by the disappearance of barbaric sports and popular recreations, but the failure of local authorities to invest in facilities, the relentlessly didactic nature of rational recreation and a firm belief on the part of the working classes that popular recreations and sports were the legitimate expression of a man's right to spend his limited leisure time as he wished, resulted in working-class sport going its own way in mid-Victorian England. Walton and Poole[60] are of the opinion that attempts to suppress popular sports were resisted by the working classes, that such sports were tolerated by employers from landed backgrounds and they conclude with Patrick Joyce that 'In the transition to a more stable and organised urban life, the legacy of older and more violent spontaneous ways was a powerful one. The 1850s and 1860s were still a time of hard-drinking, hard-sports and hard-gambling.'

The more commercial influences on leisure in the late nineteenth century – the music hall, the seaside holiday and professional sport – were to transform the popular sport occasion. Storch's cultural lobotomy[61] had thus failed to provide the working classes with appropriate motivation, and wrestling and rural sports, also known as old English, popular or rustic, were traditional activities, nearly always held on festive or national holiday occasions. Thus we find Shrove Tide sports at Milnthorpe, Easter sports at Galgate and May Day sports at Halton, Whit Monday and Tuesday sports at Skerton, Whitsuntide sports at Glasson, Trinity Sunday, Monday and Tuesday sports at Arkholme, Martinmas sports at Milnthorpe, Cowan Bridge and Burton-in-Lonsdale, Christmas sports at Marsh Lane or Boxing Day sports at Scotforth. Sports were also attached to fairs at Ingleton, race meetings at Warton, regattas at Sandside and elsewhere, and the list gives some idea of the extent of the institution whenever holiday occasions presented themselves. They were totally regular in their occurrence, and formed a major part of the recreation season of the working classes.

Malcolmson[62] concluded that by 1850, a vacuum existed in the recreations available to the mass of citizens, making them easy prey to the attractions of the publicans. While it may be true that the mass of citizens were always easy prey to the attractions of the publicans, it is less evident that a vacuum existed in the recreational life of the mass of citizens. Delves[63] believes that appreciation of the content of popular recreation is 'regrettably partial'. Yet contained within a broad, yet finely differentiated working-class social situation a wealth of activities can be identified; some are acceptable within the framework of the volunteer movement, others are on the fringe of respectability as popular sports, not for any inherent danger, but merely on account of their association with alcohol and gambling, and yet more are certainly unacceptable in the moral backlash against blood sports; finally still more are almost totally unknown among the poorest elements of the working classes but hinted at in Mayhew's *London*.[64] In the Lancaster area the range of activities covered within these popular sports events is staggering and a simple classification would include: (1) wrestling (common to most sport occasions), (2) horse races: cart-horse, trotting, hurdling, shafting, donkey, 'last-ass', (3) dog: trail-hunt, hound-trail, scent-hunt, (4) shooting: pigeon, rabbit, (5) athletics: steeplechase, foot races, tug-of-war, sack races, high jump, leaping, running high leap, high pole leaping, (6) water (where available): rowing and sailing, (7) popular (often of a skilful nature): quoits, spell and knurr, pitch and toss, (8) social: gurning, jingling, dancing.

In the debate on whether society should perhaps more readily have been divided not into social classes but into respectables and the rest, Crossick[65] has argued that the working classes established their own acceptable codes of respectable behaviour and were not coerced by the middle classes. Hill[66] equally has pointed out that the rejection of these ideals in sport often resulted in the development of a more specifically exclusive working-class form of organization, implicitly rejecting bourgeois ideology and standards.

Malcolmson's argument is nearer the mark in its claim that the failure of middle-class and middle-class-inspired institutions to attract and refine the working classes meant that the public house remained the working-class social catalysing agent. The Reverend Clay of Preston, in his social investigations during mid-Victorian England, discovered one drinking house in Preston to every 28 men. He analysed the expenditure of 131 workmen employed by one master and discovered that 22 per cent of their wages was spent on liquor, 15 of them spent up to 25 per cent of their earnings on drink and 41 spent between 25 per cent and 75 per cent.[67] Publicans changed their wares as fashions changed and there is massive evidence of the connection of popular sport with public houses. Detailed scrutiny of local newspaper reports during this period has identified 16 public houses which all regularly hosted sporting occasions for the working classes.

In addition to holiday and public house popular sporting occasions are references to attempts to promote athletic events *per se* – the Luneside sports (1844) and Lancaster rural sports in the Prince William Henry field (1855) are good examples, or sports in connection with established institutions such as the Lunatic Asylum (1855), the Septennial Boundary Riding (1858) and a variety of fêtes, especially the Municipal and Athletic Fête (1865).

Attempts by the middle classes to exert a restraining influence on the recreations of the working classes in connection with the excesses of blood sports and those activities associated with drink and gambling have been recorded[68] and similar attempts were made to control the popular sports, although as Walton and Poole and others have already indicated, not always with great success. That trouble could be expected on these occasions can be sensed from local press reports. In 1840 at the Hest Bank sports, no booths, tents or stalls were to be erected on the sands without the permission of the authorities; in 1844 it was reported of the Galgate sports that there were 'no fights', and in 1846 the editor of the Lancaster Gazette reported as follows on the Warton Races:[69]

These annual sports took place on Monday, but really, with every desire to preserve to the labouring classes their amusements, we can have no pleasure in recording these recurrences, seeing that they one and all end in drunkenness and debauchery.'

In 1848 he went a stage further.[70]

We understand it is intended to have some sports on the Sands at Hest Bank the week after next. The bill of fare comprises horse racing, pigeon shooting and some minor amusements. On hearing of this, we felt it our duty (knowing how often these sports are made a pretence for indulgence in drunkenness and debauchery) to make due enquiry, and we are assured by the responsible parties that everything will be conducted as to prevent, as far as possible, the assemblage of disorderly persons; with which view, wrestling and other popular sports are excluded.

Sports of this nature also came up against religious authorities. They were 'acceptable' on holiday occasions 'so that the holiday folks had full opportunity of getting rid of their time and money', but the Arkholme Sports on Trinity Sunday drew criticism,[71] the Christmas sports at Burton-in-Lonsdale at the Hen and Chicken Inn on Christmas Day were stopped by the churchwarden,[72] and by 1868 some popular sporting occasions had been brought to their knees. This situation reflected the national picture, although Kilvert[73] has drawn attention to the problem faced by isolated religious authorities in country districts. Yet these working-class sports, including wrestling, served several important functions. They gave an opportunity to the working man, and occasionally woman,[74] to participate in amateur sports during their free time. They afforded local working people pleasant social occasions, which, although plentiful to the middle-classes, were limited in number to their social inferiors. Working men were allowed the opportunity to achieve some sort of social status, since events were thoroughly reported, either by the local press or orally, and local champions emerged. Further, there was an opportunity to win prizes, either in cash or in kind, and finally they formed a bridge between the sports of the pre-industrial revolution period, maintained working-class interest in sport at a time when the middle classes had acted against many of the more brutalizing sports, and it could be argued, became the forerunner of more sophisticated athletics championships.

The view that popular sports meetings were mainly *ad hoc* affairs cannot be borne out by the Lancaster study. Events were advertised in the local press well in advance, programmes were drawn up, prizes and prize-money offered, booths, tents and stalls often set up and dances organized to follow the day's events. Some events drew competitors from long distances and by 1844 a committee had been established to organize the Luneside sports.[75] Bouchier and Metcalfe[76] have each made the point that participation in sport by the working classes was relegated to physical participation only and did not exist at executive or administrative levels. This may well have been true in both Canada and Victorian England during this period for highly organized middle-class oriented sport, but in Lancaster, the élite and upper middle classes did not concern themselves with the organisation of popular and public house affiliated sports. It must be hypothesized that that was left to the lower middle classes and the labour aristocrats. The railway network gradually increased attendances at the sports, and at Galgate in 1845, the mill closed early to allow its work-force both to participate in and to watch the village's annual sports.[77] By 1850 the editor of the *Lancaster Gazette* commented on the disappearance of barbaric sports and their replacement by new healthier pursuits.[78] In 1858 he again gave expression to sentiments on the value of athletic sports[79] and popular, rural and rustic or athletic sports became firmly established as annual events, along with the Gala Day of the archers and the rowing regatta. Clapham sports in the same year were reported to be excellently organized, giving rise to the comment that 'the Ancient games of Olympia were never better contested',[80] and by 1865 organization of popular sports was such that Snatchem's sports were adjourned by the committee following a disagreement over the rules of the competition.[81]

Attempts to identify participants in the popular sports, including wrestling, have been difficult. In pedestrian 'matches' or 'wager' events, the traditions of the sport often ensured that a runner's occupation was attached to or replaced a name. Thus in Lancaster between 1840 and 1850 it has been possible to identify J. Wright (millworker), Pigeon Billy Bousfield (innkeeper), Thomas Medcalf the 'Flying Cobbler', John Leeming (shoemaker), George Pym (shoemaker), William Irvine (weaver), 'Ostler' Bob and William McGinlay, the 'Flying Navie'. In popular sports and wrestling a winner's occupation is occasionally given, but scrutiny of lists of names from the 1858 Lancaster wrestling championships, have thrown up but a limited number of identifiable names and occupations. In the 'lightweight' division, occupations have been identified as cabinet-maker, shoemaker, husbandman and cord-wainer and in the 'all-weight' division, turner, joiner, tinworker, stonemason and wheelwright. Until more time can be spent on identifying participants in these sports through census returns, rent and poll books, it must be hypothesized that those who entered popular sports and wrestling events were probably involved in manual occupations and were prepared to display themselves in public combat against all-comers, both factors which are likely to have excluded middle-class participants.[82]

V

This work has attempted to focus attention on the relationship between social stratification and participation in sport in mid-Victorian England, drawing on evidence from a local investigation. It may be valuable as an isolated study or as a piece of some future jigsaw of sport in Victorian society. Thematic treatment of this nature at regional or local level may well serve to highlight similarities and differences in patterns of sport, so that a truly national picture can emerge.

Bailey[83] has reviewed the obstacles which barred the way to social integration in sport before the 1880s, highlighting the limitations for some and possibilities for others of the availability of leisure time, levels and consistency of income, cultural tradition, and the principle of antithesis in the rationale of Victorian recreation, which called for mental or physical work to be complemented by the opposite recreational activity. In addition to these practical difficulties, many of the early sports clubs were anxious to exclude the lower orders, since they were keen to ensure that the new athleticism remained untainted by savagery and became a model of civilized social culture. The new world of amateur sport became exclusive and emphasized the strength of class discrimination.

Tables 3.2 and 3.3 summarize the findings of the study so far as participation in sport in Lancaster is concerned.

TABLE 3.2

LANCASTER POPULATION (10% SAMPLE 1861)

ASC				
1 and 2	3	4 and 5	6	Total
18.90 (56)	52.00 (154)	22.30 (66)	6.8 (20)	100

Figures in parenthesis represent numbers of people.

TABLE 3.3

DISTRIBUTION OF PARTICIPATION IN SPORT IN LANCASTER (1840–70)

Sport		1 and 2	3	4 and 5	6	Total
Archery		100.00 (36)	– (0)	– (0)	– (0)	100
Rowing		100.00 (12)	– (0)	– (0)	– (0)	100
Cricket:	Period 1	89.00 (59)	8.00 (5)	3.00 (2)	– (0)	100
	Period 2	28.00 (9)	72.00 (23)	– (0)	– (0)	100
	Period 3	76.00 (34)	24.00 (4)	– (0)	– (0)	100
Bowls		50.00 (4)	50.00 (4)	– (0)	– (0)	100
Wrestling		– (0)	89.00 (8)	11.00 (1)	– (0)	100

Figures in parentheses represent numbers of people.

For Lancaster, it would appear that the initiatives for establishing sport clubs and societies lay almost wholly with the professional middle class. Such societies drew almost exclusively on ASC 1 and 2 social categories during this period, which represented less than 20 per cent of the city population and far from attempting, as Kingsley and the Christian Socialists would have wished, to use leisure and sport as a vehicle for social harmony and fusion, very obvious methods of ensuring exclusivity were employed. Thus, in archery, rowing and cricket, the formalities of membership, subscriptions, practice, uniforms and fines militated against the inclusion of working men. The segregation of rowing events and the exclusive fixture list of the cricket club were further examples of middle-class proponents ensuring that sport remained within certain class boundaries. Evidence has been provided of how fine the lines of distinction were. Master tradesmen were acceptable, but not journeymen or mechanics!

According to Tables 3.2 and 3.3, archery and rowing reveal themselves to be totally ASC 1 dominated. Cricket during periods 1 (1848–58) and 3 (1865–78) was almost equally exclusive, although period 2 (1858–65) saw an incursion of ASC 3 members, which gave a foretaste of the challenge of lower middle-class tradesmen and clerks to the dominance of the ruling cliques. It is possible to hypothesize further that participation in organized sport stemmed not only from the sporting and physical needs of participants, but from the needs of social and business fulfilment of like socio-economic groups. Thus archery remained the preserve of the gentry and professional middle class, rowing and cricket the professional and commercial middle class, bowls was commerce-oriented, and wrestling and popular sports the domain of the skilled and semi-skilled working population.

It is equally clear that the failure of the middle classes to absorb even 'the more respectable', however we define the term, members of the working classes into sport or provide acceptable, alternative forms of recreation and facilities for the working classes as a whole, resulted in the latter retaining their social identity, and possibly expressing their cohesion, through popular sports, wrestling, rowing matches and pedestrianism. Bailey[84] has expressed the view that the defence of organized sports against infiltration from below is a good example of middle-class determination to maintain existing boundaries, but he fails to recognize the effect this, in turn, had on working-class recreational identity. Equally clear from a study of Tables 3.2 and 3.3 is the fact that infiltration was not only defended within ASC boundaries 1–3, but may well have been defended between ASC categories 3–5. The failure of categories ASC 4 and 5 to make any impression on the figures for participation in organized sport confirms that for many semi- or unskilled working men, recreation may well have been centred on the public house. The incidence ot fines for drunkenness, card playing and dice gambling reported in the local press during this period may well be a reflection of the chief leisure interests.

Metcalfe[85] and Mason[86] have re-affirmed the sport historian's problem in attempting to trace, from original sources, the sporting activities of working-class Victorians. They left little documentary material, and their activities have often been reported and interpreted by socially distant, often hostile, middle-class media. Yet the perseverance of popular sports reflects how little the working classes were deterred by explicit discouragement, and how spirited was their resistance to the cultural promptings of their social superiors. From the evidence presented, in Lancaster at least, there was a wealth of popular sporting activity between 1840 and 1870 which would appear to refute Malcolmson's conclusion that it had virtually disappeared by 1850. Indeed Davis's[87] work has produced similar, if not weightier evidence of the existence of the institution in Worcestershire during the same period. There is certainly support for the view that 'barbaric' sports had all but disappeared, but participant sports available to the working classes, whenever leisure and the economy allowed, were not only regular, but had elements of organization, structure and formality which previously might hardly have been imagined.

During the period 1840–70, the view that sport represented an opportunity to achieve harmony between, and the fusion of, different social groups can hardly be supported. There is, however, evidence of a growing interest in the recreational needs of the working classes,[87] and the period 1870–1914 in Victorian society was to see progress in this respect, accelerated by the growing availability of leisure time, an improved economy, an awareness of the health and recreational needs of urban communities, better public facilities[88] and the stimulation of professional sport. Between 1840 and 1870 the view of Victorian society expressed by Disraeli, as one of 'two nations, between whom there was no intercourse and no sympathy' was particularly appropriate in sport, and class conciliation through sport remained fantasy rather than fact.

NOTES

1. See in particular P. Bailey, *Leisure and Class in Victorian England* (London, 1978), H. Cunningham, *Leisure in the Industrial Revolution* (New York, 1980) Chapter 4, and R.D. Storch, *Popular Culture and Custom in nineteenth-century England* (London, 1982).
2. W.J. Baker, 'The State of British Sport History', *Journal of Sport History*, 10, (Spring 1983), 54.
3. A. Kruger, *'Local History and Sport in Lower Saxony : a major West German project'* (HISPA Conference paper, Glasgow, July 1985).
4. R.J. Davis, 'The Development of Physical Recreation in Worcester and District, with particular reference to the last quarter of the 19th Century' (unpublished M.Ed. dissertation, University of Manchester, 1977), p.2.

5. R. Rees, 'The Development of Physical Recreation in Liverpool during the 19th Century' (unpublished MA dissertation, University of Liverpool, 1968).
6. H.E. Meller, *Leisure and the Changing City 1870–1914* (London, 1976).
7. D.D. Molyneux, 'The Development of Physical Recreation in the Birmingham District 1871–1892' (unpublished MA dissertation, University of Birmingham, 1957).
8. R.J. Davis in fact is in the process of carrying out such investigations for 15 towns and cities in the West Midlands region of England. For details see R.J. Davis, 'Patterns of Physical Recreation in different types of Urban Community in the second half of the 19th Century with particular reference to the West Midlands of England' (HISPA Conference paper, Glasgow, July 1985).
9. R.Q. Gray, *The Labour Aristocracy in Victorian England 1850–1914* (Oxford, 1976), pp. 7–8.
10. J.K. Walton and J. Walvin, *Leisure in Britain* (Manchester, 1983), p.2.
11. T. Mason, *Association Football and English Society 1863–1915* (Sussex, 1980).
12. H. Cunningham, *The Volunteer Force: A Social and Political History 1859–1908* (London, 1975).
13. W.F. Mandle, 'The Professional Cricketer in England in the 19th Century' *Labour History* 23 (1972), 1–16.
14. H.P. Mawdsley, 'The Development of Athletics in the Potteries 1850–1900' (unpublished M.Ed. dissertation, University of Manchester, 1977).
15. Gray, op. cit., pp. 99–115.
16. A. Metcalfe, 'Organised Sport and Social Stratification in Montreal: 1840–1901' in R.S. Gruneau and J.G. Albinson (eds.), *Canadian Sport: Sociological Perspectives* (Ontario, 1976).
17. Nancy B. Bouchier, 'Social class and organised sport in 19th-century Ontario: a case study of sport in a small town: Ingersoll, Ontario, 1860–1894' (paper delivered at HISPA Conference, Glasgow, July 1985). See also unpublished Master's Thesis of same title, University of Western Ontario, 1982.
18. Rees, op. cit.
19. Molyneux, op. cit.
20. Davis, op. cit.
21. Margaret H. Elsworth, 'The Development of Physical Education in the 19th Century in Blackpool and Southport' (unpublished M.Ed. dissertation, University of Manchester, 1972).
22. Using information collected by Rees, Davis and Elsworth (op. cit.) it can be seen that the period 1875–1900 witnessed a veritable explosion in club formation. The period 1840–70 was in comparison inactive, except for cricket in Liverpool.

The Explosion in Sports Clubs in Late Victorian England

	ASSOCIATION FOOTBALL			CRICKET		
	Liverpool	Southport	Worcester	Liverpool	Southport	Worcester
1860	0	0	n/a	30	5	n/a
1870	0	0	1	148	20	20
1880	4	2	14	146	47	64
1890	203	67	43	216	53	32

23. Carol Owen, *Social Stratification* (London, 1968), pp. 3–16.
24. R.D. Storch, 'The Problem of Working-class Leisure. Some Roots of Middle-class Moral Reform in the Industrial North: 1825–50' in A.P. Donajgrodski (ed.) *Social Control in 19th-Century Britain* (London, 1977), pp. 138–62.

25. J. Heiton, *The Castes of Edinburgh*, 2nd ed., 1860.
26. Gray, op. cit., pp. 99–115.
27. M. Anderson, 'Family Structure in Nineteenth-Century Lancashire' (unpublished PhD dissertation, University of Cambridge, 1969).
28. A. Armstrong 'The Social Structure of York 1841–1851' (unpublished PhD dissertation, University of Birmingham, 1967).
29. P.J. Gooderson, 'The Social and Economic History of Lancaster 1790–1914' (unpublished PhD dissertation, University of Lancaster, 1975).
30. T. Veblen, *The Theory of the Leisure Class* (New York, 1953).
31. A. Delves, 'Popular recreation and social conflict in Derby 1800–1850' in E. and S. Yeo (eds.), *Popular Culture and Class Conflict 1590–1914* (Brighton, 1981), p.101.
32.

No.	Ni.*	ASC					
		1	*2*	*3*	*4*	*5*	*6*
49	13	36	0	0	0	0	0

*Ni. = Not included, owing to lack of information.

33. See Rules and Regulations of the Society of John O'Gaunt's Bowmen (Lancaster, 1869), held in the reference section, Lancaster Library. See especially articles 1–4 and 7.
34. Bouchier, op. cit. (1985), p.7.
35. See Lancaster Rowing Club minute book (hereafter LRC minute book) held in the reference section, Lancaster Library, entry dated 20 September 1842.
36.

No.	Ni.	ASC					
		1	*2*	*3*	*4*	*5*	*6*
12	0	8	4	0	0	0	0

37. See poster advertising Chester Regatta (1846) in Lancaster Rowing Club files, held in ref. section of Lancaster Library. Depicted here as Figure 3.1.
38. *Bell's Life*, 11 October 1846.
39. See in particular, Cunningham, op. cit., pp. 128–9 and M. Girouard, *Return to Camelot* (London, 1981), pp. 240–48.
40. Pomona Gardens in Cornbrook, Hulme, Manchester, had all the usual attractions of public gardens and was frequented by large numbers of the labouring classes of Manchester and its vicinity. In summer there were frequent firework displays, bombardments, fêtes and galas. See Bradshaw's Guide to Manchester (Manchester, 1857).
41. See LRC minute book, op. cit. entry dated 30 March 1846.
42. Ibid, entry dated 17 September 1845.
43. Ibid, entry dated 13 August 1841.
44. J.J. Gilchrist, *The Lancaster Cricket Club*, 1841–1909 (Lancaster, 1910).

45.

No.	Ni.	ASC					
		1	*2*	*3*	*4*	*5*	*6*
71	3						
		37	22	5	2	0	0

46. Gilchrist, op. cit., p. 7.
47. *Lancaster Gazette* (hereafter L.G.), 28 April 1855, p. 1, c. 4.
48. Subscriptions at the time were 10s 6d.
49. Gilchrist, op. cit., p. 9.
50. In Kendal, some 20 miles to the north, the Carpet Works Club and other industrial clubs were already organized and competing, but there was no comparable development in Lancaster.
51.

No.	Ni.	ASC					
		1	*2*	*3*	*4*	*5*	*6*
38	5						
		2	7	23	0	0	0

52.

No.	Ni.	ASC					
		1	*2*	*3*	*4*	*5*	*6*
54	9						
		22	12	11	0	0	0

53. J.J. Gilchrist, op. cit., p. 78. See entry on L.C.C. v Duke of Lancaster's Yeomanry Cavalry, 11 May 1869. Note batting order of Yeomanry Cavalry.
    1. Ellesmere, Earl of
    2. Ashton, H.
    3. Part, C.T.
    4. Egerton, Honourable A.F.
    5. Fazackerley, J.N.
    6. Royds, C.M.
    7. Molyneux, C.R.
    8. Grey de Wilton, Lord
    9. Private Langdale
    10. Private Challoner
    11. Private Higginbotton
54. *L.G.* 24 July 1858, p. 4, c. 5: fees for entry to the tournament at Cockerham were 2s 6d. *L.G.* 4 October 1865, p. 1, c. 4: reports a prize of £20.00.
55.

No.	Ni.	ASC					
		1	*2*	*3*	*4*	*5*	*6*
10	2						
		0	4	4	0	0	0

56. See Owen, op. cit. (1968), pp. 20–24.
57. Storch, 'The problem of working-class leisure'.
58. W. Vamplew, 'Sport and Industrialisation: An Economic Interpretation of the Changes in Popular Sports in Nineteenth-century England', *Working Papers in Economic History* No.9, July 1985, Discipline of Economic History, Flinders University of South Australia.
59. *L.G.* 1 August 1840.
60. J.K. Walton, 'The Lancashire Wakes in the Nineteenth Century' in R.D. Storch (ed.), *Popular Culture and Custom in 19th Century England* (London, 1982), p. 103.
61. Storch, *Popular Culture*, p. 139.
62. R.W. Malcolmson, *Popular Recreations in English Society 1700–1850* Cambridge (1973).
63. A. Delves, 'Popular Recreation and Social Conflict in Derby (1800–1850)' in E. and S. Yeo (eds.), op. cit., p. 95.
64. P. Quennell (ed.), *Mayhew's London* by Henry Mayhew (London, 1984), pp. 39–46.
65. G. Crossick, *An Artisan Elite in Victorian Society: Kentish London, 1840–70* (London, 1978), pp. 252–4.
66. D.J. Hill, 'The Growth of Working-Class Sport in Lancashire 1870–1914' (unpublished MA dissertation, University of Lancaster, 1975).
67. See Gregg International, *The Working Classes in the Victorian Age, Vol. III, Urban Conditions 1848–1868* (Farnborough, 1973), p. 102.
68. In addition to the work of Storch, op. cit., E. and S. Yeo, op. cit., Donajgrodzski, op. cit., and J.K. Walton and R. Poole, see also A. Howkins, 'The Taming of Whitsun' in E. and S. Yeo, op. cit., pp. 187–208.
69. *L.G.*, 2 August 1846, p. 3, c. 4.
70. Ibid, 5 August 1848, p. 3, c. 2.
71. Ibid, 8 June 1844, p. 3, c. 3.
72. Ibid, 2 January 1847, p. 3, c. 2.
73. Rev. F. Kilvert, *Kilvert's Diary* (1870–1879) ed. William Plomer (Harmondsworth, 1984), p. 281. See entry dated Wednesday 21 October 1874.

> John Hatherell said he remembered playing football with the men on Sunday evenings when he was a big boy, and the Revd. Samuel Ashe, the Rector, trying to stop the Sunday football playing. He would get hold of the ball and whip his knife into the bladder, but there was another bladder blown up the next minute. 'Well,' said the Rector in despair, 'it must go on.'

74. There are several references in the popular sports to women's contests, and much evidence of female participation, and competition, in middle-class organized sports, in particular archery, rowing and skating. By 1870 in Lancaster there was a proposal to establish a ladies' swimming bath, occasioned no doubt by the popularity of swimming and time restrictions imposed on women in the baths. The swimming baths returns for 24 June 1865 indicated that of the 309 swimmers using the facility that week, 150 were men (1st class 65, 2nd class 85) and 86 women (1st class 27, 2nd class 59). There were further references to special footraces for women for prizes as part of the popular sports.
75. *L.G.*, 20 July 1844 and 3 August 1844.
76. Bouchier, op. cit. p. 6.
77. *L.G.*, 3 April 1845.
78. Ibid., 9 February 1850.
79. Ibid., 7 August 1858.
80. Ibid., 9 October 1858.
81. Snatchem's was, and remains, the name of a local inn by the River Lune which hosted popular sports. It was reputedly the site for press-ganging sailors, hence the name.

82.

No.	Ni.	ASC					
		1	*2*	*3*	*4*	*5*	*6*
9	0						
		0	0	8	1	0	0

83. Bailey, op. cit. pp. 124–46.
84. Ibid., p. 174.
85. A. Metcalfe 'Organised Sport in the Mining Communities of South Northumberland 1800–1889', *Victorian Studies*, Vol. 25, No. 4, (Summer 1982), 469.
86. A. Mason, 'Football and the Workers in England 1880–1914' in R. Cashman and M. McKernan (eds.), *Sport: Money, Morality and the Media* (University of Queensland, 1979), p. 248.
87. Davis, op. cit. See map sections on local sports in appendices.
88. *L.G.*, 19 February 1870.

> J. Grant, speaking at the Lancaster Mechanics Institute soirée at the Town Hall on February 15th 1870, concluded his speech by drawing attention to the recreation deprivations suffered by some Lancaster citizens.
> 'But what is the case with the toiling thousands around us? They have not their garden, or their croquet lawn, or archery field to resort to. They cannot pay their guinea subscription to the Cricket or Boat club; they cannot take a day's shooting or fishing, or a week's excursion from home at will. We must try to help them to substitutes.'
> He argued a case for the city acquiring land for recreational provision, 'as a place for cricket, football and other games . . . Ladies and Gentlemen, be determined that, through your representatives, the Town Council, you will obtain entire and permanent possession of these invaluable spots' (Applause).

4

Football and the Urban Way of Life in Nineteenth-Century Britain

R.J. Holt

We have learned a great deal about the history of our national winter sport in the last decade. Tony Mason's careful research supplemented by the work of Charles Korr and Wray Vamplew among others has provided both a coherent picture of the organizational and commercial structure of football as well as valuable data on the wages and conditions of professional footballers. Finding out who went to football matches is a difficult business because of lack of contemporary survey material. However, Mason's broad picture of a mixed male working-class crowd has received support in N.F.B. Fishwick's recent doctoral thesis on 'Association Football and English Social Life 1915–1950', which happily breaks the mould of predominantly Victorian and Edwardian research. Further notable contributions have come from Stephen Wagg whose *Football World* is useful on the 'managerial revolution' and the role of the media, and from Bill Murray's account of the sectarian history of the game in Glasgow. *The Old Firm* controversy is perhaps especially interesting in the sense that it forms a bridge from history to the kind of concerns which have preoccupied contemporary students of football as a social activity. Current concern over youthful violence at football matches has prompted some social psychologists, as well as sociologists, to try to find out if such behaviour has been a constant element over time which has been magnified out of all proportion by the media; others, however, have stressed the discontinuity in contemporary behaviour resulting from the break up of close-knit working-class communities since the Second World War.[1]

It is not my purpose to enter directly into these controversies here. Suffice it to say that, thus far, historians have quite understandably been mainly concerned with establishing the empirical framework of socio-economic change, while sociologists (with the important exception of Dunning and his co-researchers) have tended to focus on the kind of contemporary cultural meanings that have been attached to the game by its supporters.[2] This is hardly surprising given the traditions of the two academic disciplines and the kind of evidence at their disposal. Historians, in particular, have great problems in reconstructing popular attitudes. Working people rarely wrote down what they felt and the great mass of sporting journalism is notably insensitive to such questions. However, if we want to move the history of football – and of sport in general – on to new and fertile areas we have to go beyond accounts of the game couched mainly in terms of our present anxieties, whether they be about violence on the terraces or declining attendances. If we confine ourselves to the history of football as a commercial activity with 'a product to sell', we leave out a whole realm of experience to do with the playing of the game by those who had to pay for a pitch and a referee out of their own pocket. The *playing* of football has been rather overlooked in the rush to explain its commercialization. Amateurism may have been a middle-class code but we should not forget that the great majority of 'amateurs' were working-class. How then can we frame questions which draw on their experience of the game and can then be related to the wider questions which social historians have been asking recently about the nature of urban life?

Behind this study lies the conviction that, in terms of Marxist theory, an analysis which focuses discussion on the involvement of capitalists and the problems of class-consciousness cannot take us much further. Marxism has been more at home in 'work' than in 'play', although recent debates on the lines of a negotiation between dominant bourgeois values and popular resistance to cultural hegemony have been more helpful.[3] Indeed this chapter is very much concerned with the kinds of identity workers created for themselves through sport. But crude ideological manipulation through football is hard to pin down: if football was an opiate, it was a democratic one – of the people, by the people, for the people. The game was certainly not a profitable enterprise for most directors. It was hardly foisted on an unwilling public and was never commercially promoted or advertised to any significant extent. Nor is the broad alternative perspective derived ultimately from Weber more helpful in trying to pin down the essence of modern sport. The rules and structure of modern football may be both rational and bureaucratic, yet the playing of the game itself, the intense local loyalties, the factionalism and collective hysteria sit uneasily with the idea of individualism and the triumph of a scientific world view. While the search for a single core of meaning is in itself a philosophically dubious enterprise, the kind of emphasis on the way communities cohere found in Durkheim seems to offer the most fruitful opening for the researcher seeking guidance from the holy trinity of nineteenth-century social theorists. It is not so much Durkheim's rules of method as his 'urban agenda' which is broadly relevant here. Above all, he was concerned with how solidarity is sustained as the division of labour and urbanization create a potential climate of social isolation ('anomie'). This could lead to social disintegration, and in extreme cases to suicide, if there were not 'a whole series of secondary groups near enough to the individuals to attract them strongly in their sphere of action and drag them, in this way, into the general torrent of social life'.[4] Durkheim himself considered occupational associations, latter-day guilds where men would work and play together, as the best means of achieving such ends and paid no attention to the world of sport. However, if we look at the ordinary social lives of workers in the cities of late nineteenth-century Britain, we find sociable drinking and sporting institutions or associations for men were the predominant forms of communal life in a society where segregation of the sexes remained the rule rather than the exception.

In this connection it is worth taking a look at what historians of urban America have been doing. Distinctive ethnic patterns of settlement have encouraged an academic tradition of urban sociology which historians of the United States have drawn upon to produce a number of valuable accounts of the role of sport as an element in the 'search for order' in the industrial city. Sport has been a building block in the creation of stable ethnic sub-communities. It seems natural for American researchers to look at the sectional solidarities that existed within the city as well as at the role of a national sport like baseball as source of wider urban integration.[5] Perhaps it is worth taking a similar perspective on Britain despite the relative lack of racial and linguistic diversity. This certainly ties in with the shift in social history as a whole towards a better understanding of popular attitudes, and the emphasis of sociologists of post-war football like Critcher, Robbins and Taylor on the disintegration of established working-class communities and culture.[6] There is plenty of unacknowledged common ground here between social history and contemporary cultural analysis in the sense of the need to see sport from 'below' by trying to reconstruct some of the meanings ordinary people attach to their own experience. How does the rise of sport fit into the idea of community in the industrial city?

II

This is a huge subject and my intention here is simply to raise a number of issues that seem to be worth following up in order to encourage further research. For the moment some of these topics can only be the subject of speculation, and much of what follows amounts to little more than the setting out of hypotheses. So much for the broad theoretical context, what about the specific issues?

Three related questions spring to mind: first, how far should we see football not as an invention but rather as a form of cultural continuity, especially as far as the traditions of male youth are concerned? Perhaps we have taken on board too eagerly the heroic accounts of the public school men, who founded the Football Association in 1863, and assumed in consequence that traditional football was suppressed lock, stock and barrel during the first half of the nineteenth century to be re-invented and re-popularized in the second.[7] Concern for underlying continuities in popular behaviour leads us to consider the social functions of the new clubs themselves and prompts a second question: how did the modern organization of football develop at the grass-roots level? What were the links between the daily life of street and pub and the world of football? Third, why did professional football become so powerful a vehicle for collective urban identity? What submerged festive role did football come to play in the creation of a stable working-class culture?

So, from a variety of directions – from evident gaps in our knowledge as well as from considerations of interdisciplinary convergence and theoretical orientation – the question of continuity and adaptation in sport needs to be examined. Unfortunately there is no detailed study of urban sport which spans the whole of the nineteenth century. We have inherited a way of looking at things which sees sport very much in terms of traditional and modern forms, although recent work on the eighteenth century in particular has shown the shortcomings of so simple a distinction. 'The commercialization of leisure' announced by Plumb and elaborated recently by Brailsford stresses the vigour of sport in the late eighteenth century.[8] Malcolmson's sensitive account of the undoubted success of humanitarian reformers in putting down cock-fighting and bull-baiting in the second quarter of the nineteenth century has been confused with a decline of sport more generally. This idea of a leisure vacuum is enhanced by well-documented accounts of the suppression of the great festive football matches such as that offered by Delves for Derby.[9] The trouble is that the banning of brutal sports, and of the most notorious of the ancient football games, tends to obscure much of the day-to-day sporting life of the time. Cunningham has made this point forcefully, but only Metcalfe's recent study of the miners of south Northumberland has clearly identified the problem. However, most of the evidence he draws upon comes from the later rather than

GRADIDGE,

Manufacturer of the

FAMOUS "INTERNATIONAL" FOOTBALL.

9/6
Each.
OR
Buttonless
(as Sketch)
10/-
Each,
Post Free.

9/6
Each,
OR
Buttonless
(as Sketch)
10/-
Each,
Post Free.

Used by all the leading Clubs, Schools and
Colleges in the Kingdom.

ADDRESS :—

FACTORY,

ARTILLERY PLACE,

WOOLWICH, S.E.

2.   A vital part of the burgeoning football industry (Referees' Association Chart,
1896–7)

"Well, even if 'e did cost twelve thousand, 'e doesn't know 'ow to
pass the ball to the wing."
"Don't talk silly. D' ye expect 'im to pass the ball to a man wot only
cost twelve 'undred?"

3.   A fan's view of the transfer system (from *Mr Punch's Sports and Pastimes*)

BENETFINK & Co.,

THE GREAT CITY DEPOT FOR COMPLETE

Football, Hockey and Lacrosse Outfits.

CATALOGUES POST FREE.

FOOTBALL SHIRTS.

	Boys'	Men's
Plain Flannelette	2/3	2/6
Harlequin Halves or Quarters do.	2/6	2/9
Strip'd Flannelette	2/6	2/9
Flannel, all Wool any colour ...	4/3	5/-
Harlequin Halves or Quarters do.	4/6	5/3
Plain Navy Melton any colour ...	—	2/6
Ditto Trimmed Sateen... ...	—	3/-

PATTERNS POST FREE.

N.B.—Allowance on taking 1 doz. Shirts, 3/-.

POSTAGE—1 Shirt, 3d.

FOOTBALL BOOTS.

Benetfink's "Premier," Best Calf, Double Ankle Pads. Boys', 6/11; Men's, 7/11.

Benetfink's "Universal," Best Russet Hide, Single Ankle Pad. Boys', 5/11; Men's. 6/11.

Benetfink's "Defender," Best Chrome, Latest Pattern. Men's. 9/11.

POSTAGE—Single Pair, 6d.

FOOTBALL KNICKERS.

	Boy's		Men's	
White Swansdown ...			1/4	
White Lambskin	2/-	2/8	2/2	2/10
Blue Serge	1/9	2/4	1/10	2/6
Do. all Wool	3/3	4/4	3/11	4/11
Best ditto ...			5/11	

PATTERNS POST FREE.

N.B.—Allowance on taking 1 doz. Knickers. 2/-.

POSTAGE—Single Pair, 3d.

BENETFINK'S

"1st Practice," full size, 5/10 net. Post Free.

THOMLINSON'S

Eight Panel Match Ball Post Free, 8/6.

PATTERNS POST FREE.

All Orders over 10/- sent Carriage Free to any address in the United Kingdom.

89, 90, 107, 108, CHEAPSIDE, LONDON, E.C.

4.   Distinctive playing outfits promoted team identification and loyalty (Referees' Association Chart, 1896–97)

from the early-to-mid nineteenth century and he concentrates his attention on bowling and quoits rather than on football.[10]

Early accounts of the history of association football, or of rugby football for that matter, ask us to accept that these games were passed down to the urban workers by those who had learned them at public school. While no-one is disputing the key role of élite education in the drawing up of new rules for football, we should not therefore take it as read that the older forms were moribund. We cannot simply assume that so long and well-established a folk game had simply disappeared by the middle of the century. We know from the research of the Opies on children's games how tenacious are the traditions of the street.[11] Games were passed on and adapted by word of mouth and by example from generation to generation of children despite unprecedented urban and industrial upheaval. The way football continued to be played 'in courts and alleys, on vacant plots of land, on brickfields, indeed on any open space at all that may be found' reminds us that casual football with hazy rules and imaginary touchlines continued to be a part of childhood.[12] Stephen Humphries has given an account of the battle with the police to stop street football in the Edwardian period viewed from the perspective of the participants themselves. Kicking a rag and paper ball round the streets in the early twentieth century may not have been so different from the way boys had played in previous generations and might be considered as much a survival as an innovation copied from their betters.[13]

We also know that as association football spread in the later nineteenth century it was not played by the working class in the same spirit as in the public schools. Much of the intense localism of the ancient forms of football survived in the new game and in the Northern Union game of rugby.[14] Moreover the phenomenal speed at which football spread from the mid-1870s to the mid-1880s in cities like Liverpool and Manchester rules out the possibility of seeing the game simply as a rational recreation initiative 'from above'. Nor does it seem sensible to read into this growth only the copying of middle class games, or to put it down to the influence of the press as coverage followed rather than preceded the spread of the game. The conventional explanation which is given, for example, by Molyneux for the growth from around 12 to 155 clubs in Birmingham between 1875 and 1880 is that reduced hours of work now permitted more play.[15] But that begs the question why did workers choose *this* particular form of play? Perhaps the reorganization of the working week, which careful research has shown did not mean a reduction in hours so much as a more even pattern of labour, simply facilitated the formalizing of well established traditions of street football into regular time slots and new organizational structures? How do we know that boys and men, especially apprentices with their corporate rites and traditions, did not carry on playing 'football' until the opportunity to convert their enthusiasm into a regular and codified league structure came along? We know that the big occasions for traditional football were largely suppressed, but we know next to nothing about the more casual playing of the game. Only a meticulous ethnography of the mid-Victorian urban working class will give us some clues. Certainly, if evidence of the continued popularity of football on any scale can be found before the arrival of the association system, the rapidity with which the northern working-class teams took over the sport and their proprietory attitude towards it becomes more readily comprehensible. In their own eyes they may simply have been asserting ownership over what had been theirs all along. It may be that there is no evidence and that such questions cannot be answered. But at least we should try. Doctoral study at the local urban and rural level over the nineteenth century as a whole using the local press from its inception seems desirable. Work under way in Scotland along these lines certainly seems to suggest that old-established sports do not seem to have declined in the lowlands and central belt in the earlier part of the nineteenth century.[16]

III

We are on rather firmer ground when we come to look at the second question concerning the relationship between the organization of teams and the actual experience of urbanization. Anderson's detailed research on the census returns of 1851 in Preston – a classic town of the Industrial Revolution – reveals the continued importance of kinship and of other relationships which often originated in migration from the same village to the city. Most migrants came from relatively close by, 40 per cent living less than ten miles from their place of origin and only 30 per cent more than 30 miles away.[17] Life was less a matter of the anonymity of the city than the intimacy of the street. In a survey of Sheffield in 1889 the Medical Officer of Health remarked that the city 'more closely resembles a village than a town, for over wide areas each person appears to be acquainted with every other, and to be interested in that other's concerns.'[18] Local networks sprang up and a person might often know a large number of others by name and even more by face. The immediacy of the 'urban village' struck middle-class observers like Masterman who noted in 1904 that a working man '"dumped" down in some casual street, unknown to his neighbours, unconnected with a corporate body or fellowship . . . goes through his life in a kind of twilight, dimly wondering what it all means.'[19] Of course, such communities had their own complicated hierarchies of rank and respectability. The working class was not an undifferentiated sociable mass, and this further served to emphasize the overwhelming importance of the locality and the street. Long streets even had their 'rough' and 'respectable' ends, with subtle gradations in between, only properly understood by the residents. What side or bit of the street you lived in could be most important in the 'microsociety' of the city.

Organized team sports, especially football and cricket, were integrated into this close-knit pattern of collective life through neighbourly solidarities and the 'local'. This was street corner society with its banter and 'characters', its 'hard men' and bookies' runners, who surreptitiously plied a universal trade with the punters. Such was the street life of Glasgow, that most avid of footballing cities, and it would be interesting to know more about the way boys moved from the back lane 'kickabout' into the more organized world of club football. What seems to be clear is that such teams often sprang from a formalizing of casual street relationships, bringing a shape and a continuity to that most basic of feelings – the sense of place. City boys were cut off from the countryside as the suburbs spread out and, lacking proper parks and playgrounds, they simply took over the streets, kicking a ball around in between playing dozens of other games. Organized teams provided a bridge between the childish and the adult male world, a means by which the playful enthusiasm of boys was turned into the tougher style of men. Here boys learned how to drink and tell jokes as well as the language of physical aggression. Sport was a part of this process. You had to learn to 'put yourself about', to 'let him know you're there'. The talents of 'tanner-ball' players, whose skills had been refined with a kid's rubber ball, were all very well, but durability was the most basic requirement. Playing football called for a strong kick and plenty of puff. Football clubs were only a part of the wider process of male socialization, which took in the workplace, the pub and the world of hobbies. However, it would be very interesting to know how sport overlapped with other better-known masculine institutions. To get direct and detailed answers to such questions an oral history project would be required, which would necessarily have to concentrate on the inter-war period and after. However, by the creative use of the census enumerators' books in conjunction with team lists, the occupational, residential and family structure of late nineteenth-century clubs might be reconstructed at least in part. This kind of urban historical anthropology may be the new frontier of sports history.

For the moment we can only point to the fact that the naming of teams seems to bear witness to the close correlation between clubs and locality, particularly the street. That clubs normally named themselves after a specific place seems too obvious to mention until we think how easy it would have been for clubs to be named on the basis of factory, political or religious affiliation, place of entertainment or after important local people. Church teams, of course, were quite prominent, although these may well have concealed older territorial loyalties to the parish beneath the cloak of religion. Church teams in Sheffield declined when a change of ecclesiastical policy required their players actually to attend services in 1925.[20] Works-named teams only accounted for 20 out of 218 football teams mentioned in the Birmingham press between 1876 and 1884 and only 25 out of 214 cricket teams.[21] Judging by names – and this is admittedly only a very rough and ready means of identifying the actual origin of clubs – neighbourhood relationships counted for more than the solidarity of factory or church. Around half the Liverpool cricket clubs examined by Rees had street or place names.[22] Tony Mason notes that there were a whole string of clubs in Blackburn that seem to have grown up on the basis of formalizing street corner sides: there were, for example, Red Row Star, Cleaver Street Rovers, Gibraltar Street Rovers and even a George Street West Rovers.[23] In South Wales rugby developed not only in the mining valleys, where playing together was an expression of the special solidarity of pitmen, but also in the larger coastal cities like Cardiff and Swansea. Here a similar pattern of residential solidarity is suggested by Smith and Williams. Swansea had its Goat Street and East Side Rovers while Llanelli had Gower Road and Prospect Place Rovers.[24] Research by Dr N.L. Tranter into sport in nineteenth-century Scotland has identified 68 football teams in Stirling (excluding the surrounding district) formed between 1876 and 1895 – a very substantial figure for a town of around 17,000. Of these 68 teams, 37 were named after a specific neighbourhood of the town. Five sides explicitly took street names: George Street, Baker Street Rangers, Shore Road Thistle, Wallace Street Thistle and Cowane Street Thistle. But close analysis reveals other sides of street origin, for example, 'Our Boys' and 'Stirling Hawthorn', both of which were described in the press as 'Cowane Street' teams.[25]

The predominance of the street or immediate locality as the basic unit of sporting organization bound together older civic traditions of living 'in public' (i.e. buying and selling, talking and playing in the street) with the particular ties of family and friends created by migration from the country to the city. Moreover, the pattern of industrial and urban change in the late nineteenth century tended to accentuate residential rather than workshop solidarities. This came about as a result both of the expansion of cities and the growing size of factories. It became increasingly less common to live and work in the same locality. Although it would be mistaken to underestimate the power of close-knit trade communities in inner London or in mining and shipbuilding centres, as a whole the spread of the rail network combined with buses and bicycles encouraged commuting. The London County Council estimated that 820,000 manual workers were making extensive journeys to work by 1905.[26] The point here is that a man was more and more likely to find himself living alongside others in different trades than before. Men no longer lived 'over the shop'. Big company towns like Crewe or Middlesborough were exceptions to this, though even there the scale of operations was such that a man could never know more than a fraction of his work mates. The large mechanized factory became the standard national unit of economic organization in the late nineteenth century, whereas the concentration of hundreds or thousands of workers together had been exceptional outside of the textile sector before the 1860s.

Men probably had their 'mates' at work and their 'mates' at home. A tradesman could no longer be sure of finding a workmate at the local with whom he could talk shop. The gossip of the older craft workshop that had naturally carried over into the alehouse nearby might be out of place in a larger public house miles away. 'The best thing you can say for football', wrote an observer of Glasgow in 1901, 'is that it has given the working man a subject for conversation.'[27] Sport, of course, had always played an important part in the traditions of recreation and sociability associated with drinking. In early modern England alehouse keepers organized sports in adjoining fields or in their courtyards and backrooms. Prominent tavern-keepers had been important in the running of bigger sporting events in the eighteenth and early nineteenth centuries, especially prize-fighting and rowing matches where they were often appointed the holders of a large purse.[28] Publicans were also involved in the provision of private running grounds for pedestrianism in the mid-Victorian period. These traditions lived on in boxing and in football. Taking over a pub was the preferred choice of occupation for successful professional sportsmen, and rich publicans themselves sometimes aspired to a place in the directors' box. Stan Shipley's study of Tom Causer of Bermondsey, a leading boxer of the 1890s and landlord of the Eight Bells, provides an important insight into the sporting dimension of pubs, some of which put on boxing for their customers from time to time.[29]

As far as football goes, Tony Mason makes the distinction between 'the use of pubs by football clubs and the establishment of football clubs by groups of people frequenting the same pub'. Before municipal facilities were provided, pubs were the obvious places for teams to meet and change before a game. Some pubs even provided a field for the game as the White Hart did for Tottenham Hotspur. Newton Heath began life changing in the Three Crowns in Oldham Road before they became Manchester United and established their headquarters at the Shears Hotel.[30] Of course, if clubs became very successful their ties with a particular pub would be weakened as they became the cultural property of the whole town. But such clubs accounted for only a small minority in relation to the myriad of sides who met in their local pub to change or to chat after the match, who probably fell into Mason's latter category of 'customer' teams. Going to the pub after a game of football is still very much part of a Saturday afternoon's amateur football. These customs die hard. A few clubs were even named after their 'local'. Molyneux lists 13 of the 218 football teams identified by pub names in Birmingham between 1876 and 1884.[31] The development of pub football leagues is a subject worth further investigation and may reveal far greater pub involvement in team sports than we have yet realized.

The pub was the central social institution of the adult male working class and occupied a pivotal position in the world of sport. Pubs were places where men both played games and talked about them. They provided facilities for changing and meeting as well as sponsoring local cups on occasion, and from the mid 1880s enterprising landlords increasingly tried to boost their custom by offering a results service as well. Club secretaries were asked to wire in results of district league matches. This was soon superseded by the growth in the 1890s of the Saturday afternoon results edition of the local newspaper, known as the 'pink 'un' or 'green 'un', which meant that both local and national football news were available almost immediately for discussion in the pub or on the streets. On late Saturday afternoons in Glasgow 'at the corner the kerb is covered with men who stand with their backs to the light, intently reading in a pink newspaper: "Haw Wull, whit's the score? . . . Hi man! whit's the score? . . . Whit the —— are you shouting at?" '[32] This habit lived on until the age of television. In Barnsley the Saturday evening streets were 'a sea of "Green 'Uns"' into the 1950s and 'occasionally a man could be seen reading the "Green 'Un" behind his partner's back as they waltzed'.[33] These specialist papers combined the Football League results with local gossip, often written in dialect and assuming an intimate knowledge of an area, its characters and the foibles of referees and pitches. Fishwick's work in particular points to the importance of these special papers as a valuable source of material for future research.

IV

So football talk was a staple element of male conversation. There was always some new item of information or odd result to be discussed. This feature of what had become the most heavily industrialized and urbanized society in the world in the late nineteenth century leads us naturally from the realm of the local to the national and brings us to the third aspect of football raised at the outset: its emergence as a powerful expression of popular civic pride. All this seems so natural, so much an unchanging element of British life, that we have only begun to question the divine right of so many towns to have their own team as falling attendances, debt burdens and the prospect of closures have to be faced. With influential forces in the game demanding a shift towards a new structure based on a 'super league' of top teams that will draw a national audience of television viewers on the lines of the American Football League, the existence of so many professional sides, each with their own ground, full-time staff and a dwindling number of loyal supporters is starting to look like an anachronism; professional football as we know it may be the product of a distinct epoch in the social and cultural history of urban life. It is this distinctiveness which requires emphasis. The emergence of the Football League structure of ranked divisions has tended to be seen as an inevitable commercial outcome of the reduction in working hours and the rise in real incomes that took place in the second half of the nineteenth century. Now it may be the case that greater amounts of time and money were 'necessary conditions' for the rise of football as a modern commercial entertainment but this does not mean that material changes alone (including improved transport and the rise of a popular press) provide a 'sufficient cause' for the phenomenon. Economic changes help explain how professional sides became financially viable, but they do not explain why so many people were anxious to identify themselves emotionally with a team of players hired to represent a club, which in turn represented a town. The public-school sportsmen found such loyalties at best confusing and at worst degrading. The middle and upper classes had stronger alternative ties to school, profession or the state. Urban identity was less important to them than it was to the workers who congregated in their tens of thousands to affirm their undying devotion to Newcastle, Derby or Bolton. We know perfectly well that these vast crowds, which after all represented the largest regular gatherings in peace-time, were fiercely partisan. They did not attend football just to enjoy the game for its own sake, as a pure celebration of speed, skill, team work and strategy. Of course, there was, and is, some objective assessment of play – knowing nods and shrewd tactical asides – but crowds tend to be more generous in their praise of the opposition when the home team is on top. Those who went to football matches were not ordinary consumers of leisure. Being involved with a football team was a very different kind of experience from buying a newspaper or a book, or going to the music hall or the cinema. Although there were 'floating' attenders whose loyalty to a club was 'conditional' as Jeremy Crump points out in his recent analysis of leisure in Leicester, the bulk of the crowd were regulars.[34] It is worth thinking for a moment about the words that were used to describe them: 'supporters', for example, carried with it a sense of unshakeable loyalty, enjoying the good spells and stoically putting-up with bad patches. Calculations of climate or league standing came second to just turning up, just being there to support your team and to identify yourself with thousands of others who felt the same. Even more telling was the word 'fan', abbreviated from 'fanatic' and carrying with it the distinct whiff of moral disapproval. 'Fans' were too lacking in personal fitness, too narrowly prejudiced to be considered 'sportsmen' by their 'betters'. Despite the presence of public-school men in the upper echelons of the Football Association and in some of the boardrooms of leading clubs like Arsenal, watching football was not considered to be quite respectable before the First World War. The King did not attend the Cup Final until 1914, and even then *The Times* correspondent considered the result to be of less interest than the fact that the cuckoo had been heard calling in the vicinity of the ground.[35]

Elite incomprehension stemmed partly from the way the middle classes steadily withdrew themselves from life in the new industrial cities. Middle-class children were frequently sent away to school where they learned to play rugby while their parents moved out to leafy suburbs within commuting distance of the inner city. The city itself was left to the workers. The success of the professional football league is closely bound up with the tension between a strong urban identity and the awareness of a wider national dimension. Football provided not only an exciting game to watch but a framework within which the one industrial town had a weekly opportunity to defeat another and, over the season, to establish dominance within the nation as a whole. The early success of smaller industrial towns like Darwen soon gave way to the domination of the bigger northern industrial cities, which not only had large catchment areas to sustain professional teams but were also avid for recognition through their sporting prowess. Only two of the original twelve clubs which formed the Football League came from towns of under 80,000 and these two, Accrington and Burnley, could hardly be considered small; moreover the pace of urbanization did not slacken. There were 36 cities of over 100,000 by 1911, which alone accounted for a quarter of the total population of England. The support for professional clubs, which averaged around 16,000 for home games in the First Division of the English Football League by 1908–9, was based on a new kind of dual urban identity. On the one hand men lived in intimate neighbour-hoods where they played with their friends and relations, their neighbours and work-mates. On the other they were also citizens, members of new economic, administrative and political units of hitherto unimagined provincial scale and complexity.

With the spread of elementary education and the rise of the popular press locally and nationally, citizens were better informed and had wider horizons than before. The structure of organized labour shifted from tight craft-based unions to larger national bodies representing most forms of skilled and unskilled manual labour. The nature of the firm was similarly transformed from the small family-run factory to the larger operation dominated by managers responsible to shareholders. In the place of the kind of industrial squire, whose dominance over the mill towns of mid-Victorian Lancashire has been examined by Patrick Joyce, came the more impersonal management structure of the limited liability company of the late Victorian period. Even the large paternalist ventures ran out of steam as the reforming and evangelical impulse waned among employers.[36] Men were left more alone to run their own lives. Imperialism through which 'Englishness' found its most potent expression might have provided a system of belief for the middle and upper classes, but its impact on urban industrial workers was less clear-cut. Cricket certainly fostered an awareness of elements of the Empire and the 'home internationals' in football and rugby shored up historic national identities in new ways. Yet nation and Empire were never fully assimilated into working-class culture.[37] The city provided the immediate context for daily life and a more readily understandable framework for collective identity than the Empire. Professional football offered a second identity to the inhabitants of the cities, which went beyond the confines of the 'urban village' and gave recognition to the sheer scale of urban life in the new scheme of things. Football as a spectator sport provided a means by which men could come to terms with the reality of the late Victorian city and clarify their relationship to it. By supporting a club and assembling with thousands of others like himself a man could assert a kind of membership of the city, the heart of which was physically and emotionally his for the afternoon. The huge grounds themselves towered over the surrounding houses and could be seen from far off. As three o'clock approached and the trickle of spectators became a flood in the streets leading to the ground, the workers briefly took possession of the city.

In the largest cities, where problems of identity and integration were potentially most acute, there were very often at least two major teams which represented a fairly well demarcated part of the whole. Derby games gathered together huge numbers of citizens, often 50,000–60,000. A single club was insufficient to represent a city of over half a million. Not only could the urban economy support two professional sides, historic and geographical divisions sometimes required this. Football both reflected and reinforced territorial and cultural differences within cities. Yet by competing so fiercely for the sporting dominance of a particular city, derby games paradoxically strengthened rather than weakened civic pride. Even in the west of Scotland, where sectarian divisions found their most vehement expression in the Celtic and Rangers rivalry, the sense of pride in Glasgow was enhanced rather than diminished by the struggle. For many Glaswegians, whether Protestant or Catholic, Glasgow football was Scottish football and the result of the 'Old Firm' game was crucial, although there were several other professional teams in the city.[38] In Edinburgh and Dundee there were similar, if less fiercely sectarian, binary divisions. In Manchester, Nottingham, Birmingham, Sheffield and Liverpool the derby rivalries, in which each club came to depend upon the presence of the other, grew up. In London the sheer scale of urbanization ruled out a simple duopoly, but encouraged the formation of a more complicated pattern of community identity and competitiveness. However, football in London has been relatively neglected and deserves further study. In Sheffield the United–Wednesday match drew crowds of up to 60,000 with the two clubs dividing up the city between them. The local football special, the *Green'Un*, remarked in 1911 that 'most of the Wednesday players live close to each other, and during the week and close season are to be seen together, playing bowls, etc., mingling with local people who become attached to them and cannot see wrong with them when they play'.[39] 'Derby' rivalries were not confined to teams from within a single city. Newcastle and Sunderland games drew huge crowds and were notoriously hard-fought matches. Geordies sharpened their sense of who they were through such encounters with their neighbours, and *vice versa*. Belonging to the community of the Tyne or of the Wear was what really mattered in the north-east. Thus the Football League both reflected and reinforced divisions within cities and particular rivalries between them, and the full extent of this merits properly detailed analysis.

The importance of local rivalries points up the festive side of professional football. The popular culture of early modern Europe had been steeped in parish feuds. Amid the routine of weekly league games there were the big occasions in which some of the old pageantry of public life and the *carnavalesque* traditions of the past were adapted to new conditions. Christmas was quickly established as a kind of football festival with games on Christmas morning and Boxing Day, usually against a familiar opponent. New Year's Day games were also very popular (with the Celtic–Rangers game attracting crowds of over 50,000 in the immediate pre-war years rising to the record attendance for a league game of 118,567 on Ne'erday 1939). For big games there were the team colours to be worn, painted umbrellas, rattles and bells to be swung. When a goal was scored the home crowd would sway and leap with joy and relief. Football was one of the few arenas left where a partial loss of self-control was permissible, where even the most stolid men might let themselves go for a moment. Antics and gestures that would have been considered crude, puerile or even dangerous elsewhere were tolerated in football. Although there seems to be little evidence at this time of the organized violence of young supporters, which has been so evident at football since the late 1960s, there is no doubt that football matches were often the scene of furious swearing and the odd brawl between drunks.[40]

V

More important for the present argument, however, is the way football offered a kind of periodic affirmation of collective identity, which found its highest expression in the celebrations that would follow upon the winning of a trophy, especially the Cup or the League. For northern fans, who made up the great majority of spectators at professional football before 1914, the chance to go to London for a big game was the experience of a lifetime. Thousands of men would go together on specially hired trains, singing and shouting as they spilled out along the London streets. Even more awaited the return of the triumphant team. When a team came back with 'a bit of silver plate' the town would be given over to a huge celebration that might only be surpassed in scale by a coronation or the end of a war. When Preston won the Cup in 1889 the town took to the streets in a kind of public festival which was reminiscent of the unrestrained Wakes of the previous generations. 'Around the team swarmed hundreds of fanatics, each of whom struggled to get a handshake with a member of the team.' When the trophy was held aloft 'the wildest enthusiasm prevailed. Hats were thrown up, handkerchieves waved, and sticks flourished'.[41] Similarly when Spurs won the Cup in 1901 – the first non-League and southern club to have done so since 1882 – the crowds blew horns and chanted snatches of popular songs changing the words to suit the occasion. 'There were so many people, so tight together, you could have walked on their heads.'[42] This went on all night and a great many of the revellers returned the next evening to see an animated picture show of their success, which came after a replay with Sheffield United held in Bolton. Similar scenes of popular enthusiasm had a long history and such moments of intense collective life were probably much more common in pre-industrial society where there had been a far more elaborate calendar of public festivity. The specific focus and forms of revelry were new but its essentially local and uncommercialized character was reminiscent of earlier festivals. Whether in football, or among the rugby league clubs of Yorkshire and Lancashire, the winning of trophies brought forth a great outpouring of feeling with teams marching through the streets accompanied by a band to public acclamation. This is what happened when Halifax won 'T 'Owd Tin Pot' (the Yorkshire Cup) in 1894 and proceeded to fill it up with champagne and pass it round their supporters. Even very small towns, little more than industrial villages, could produce crowds of thousands to see the town win a district challenge cup.[43]

Football as a form of urban festivity seems worth further investigation. The continuities at the level of the small and more tightly integrated communities are very striking. However, the celebrations surrounding, for example, Tottenham's victory, which was won by a team made up of five Scotsmen, two Welshmen, an Irishman and three northerners, is even more interesting. It mattered not at all that their team were 'mercenaries', as the indignant middle-class amateurs called professional players. That none of them was local may have been a cause for only slight regret. No doubt crowds preferred players from their own area (although this sometimes brought to the surface the neighbourhood rivalries that were normally submerged in the wider following for the senior team), but the most important thing was to have a good team. Even if many of the players had no organic link with the club, they tended to live within the community and were given privileged honorary membership of it as we saw was the case with Sheffield Wednesday. Football clubs may have been owned in the legal sense by a handful of wealthy men and staffed by a shifting group of players willing to play for anyone who would pay them, but the emotional bonds that held these clubs together were real and enduring. The problem is how to discover more about football as an experience, as a kind of endless variation on a common theme of pride, courage, skill and strength. To the fans their teams embodied an idealized collective vision of themselves and their communities, 'a story people tell themselves about themselves' as an anthropologist such as Clifford Geerz might put it. This pride was a kind of unifying urban myth, punctured by the odd shaft of realism. The faithful were permitted to joke among themselves about their team while defending it stoutly against the criticism of unbelievers.

As the scale of the industrial city outstripped the capacity of individuals to compass it, the fact of being a supporter offered a sense of place, of belonging and of meaning that could never come from the formal expression of citizenship through the municipal ballot box. In the concluding sentence of his masterly study of the early development of professional football, Tony Mason remarks that 'merely watching' the game 'brought a new kind of solidarity and sociability'.[44] My purpose here has been both to expand upon that insight and to draw attention to the playing of football as well as to those who watched it. What are required now are several original local studies of sport, informed by the wider debates within social history, about the nature of urban and working-class experience. We need to know not only more about the culture of football but much more about its links with other sports. After all, many footballers played other sports too – but which ones and with whom? As far as the nineteenth century is concerned the study of the local press, especially the specialist football papers, might permit a more precise reconstruction of the organization of football as a participant sport. The social and residential composition of a wide range of sports teams might be reconstructed using census enumerators' books as well as local directories. Whether there is any evidence for the survival of the older traditions of street football feeding the newly formalized clubs is doubtful, but the distinction here between casual 'teams' and organized 'clubs' is worth following up. Even if this search for origins turns out to be fruitless, the wider question of the adaption of popular festivity and urban patriotism through professional football offers scope for the study of cultural continuities. And for anyone wishing to pursue these themes into the twentieth century, oral history gives the possibility of finding out at first hand about the feelings, relationships and attitudes of players and spectators that we can only dimly and imperfectly perceive before 1900.

NOTES

1. Tony Mason, *Association Football and English Society 1863–1915* (Brighton, 1980); C. Korr, 'West Ham United F.C. and the beginnings of professional football in East London, 1895–1914', *Journal of Contemporary History*, 2 (1978); W. Vamplew, 'The Economics of a Sports Industry: Scottish gate-money football, 1890–1914', *Economic History Review*, IV (1982), (4); N.F.B. Fishwick, 'Association Football and English Social Life' (Oxford D.Phil., 1984, unpublished); see also S. Tischler, *Footballers and Businessmen: the origins of professional soccer in England*, (New York, 1981), and S.G. Jones, 'The Economic Aspects of Association Football in England, 1918–39', *British Journal of Sports History*, I (December 1984); S. Wagg, *The Football World: A Contemporary Social History* (London, 1984), and Bill Murray, *The Old Firm: Sectarianism, Sport and Society in Scotland* (Edinburgh, 1984).

2. E. Dunning and P.J. Murphy, *Working Class Social Bonding and the Socio-genesis of Football Hooliganism*, S.S.R.C. Research Report, June 1982; for a recent review of the sociological literature see J. Pratt and M. Salter, 'A Fresh Look at Football Hooliganism'. *Leisure Studies* 3 (1984).

3. Jennifer Hargreaves (ed.), *Sport, Culture and Ideology* (London, 1982), Chs. 1 and 2; also R. Gruneau, *Class, Sports and Social Development* (Massachusetts, 1983) attempts a fusion of history and Marxist theory.

4. E. Durkheim, *The Division of Labour* (Glencoe ed., New York, 1964), p. 28; see also comments of S. Lukes, *Emile Durkheim* (London, 1975), p. 330, and A. Giddens, *Capitalism and Modern Social Theory* (London, 1971), esp. pp. 104 and 112.

5. For a good survey of this literature see M. Adelman, 'Academicians and American Athletics: a decade of progress', *Journal of Sport History* (Spring 1983); the work of S. Hardy and B. Rader is particularly relevant here.

6. I. Taylor, 'Soccer Consciousness and Soccer Hooliganism' in S. Cohen, *Images of Deviance* (London, 1971); C. Critcher, 'Football since the war' in C. Critcher, J. Clarke and R. Johnson (eds.), *Working Class Culture: Studies in History and Theory* (London, 1979); D. Robbins, *We Hate Humans* (London, 1984).

7. The 'authorized' version comes to us through A. Gibson and W. Pickford, *Association Football and the Men who made it* (London, 1906), 4 vols., and is embodied in P.M. Young, *A History of British Football* (London, 1968), and less dogmatically in J. Walvin, *The People's Game* (London, 1975), pp. 56–9; H. Cunningham, *Leisure in the Industrial Revolution* (London, 1981), esp. pp. 127–8.

8. J. Plumb, *The Commercialisation of Leisure in Eighteenth Century England*, The Stenton Lecture, University of Reading, 1972, and D. Brailsford, 'Sporting Days in Eighteenth Century England', *Journal of Sport Hist.* (Winter 1982).

9. R.W. Malcolmson, *Popular Recreations in English Society, 1700–1850* (London, 1973); A. Delves, 'Popular recreation and social conflict in Derby, 1800–1850' in S. and E. Yeo (eds.), *Popular Culture and Class Conflict, 1590–1914* (Brighton, 1981).

10. A. Metcalfe, 'Organised sport in the mining communities of South Northumberland', *Victorian Studies* 4 (1982).

11. I. and R. Opie, *Children's Games in Street and Playground* (Oxford, 1984 ed.), esp. pp. 4–14; see also J. Walvin, *A Child's World: A Social History of English Childhood 1800–1914* (London, 1982), Ch. 5.

12. Charles Russell cited in D. Rubinstein, 'Sport and the Sociologist, 1890–1914', *Brit. J. of Sports Hist.* I, 1 (May 1984), 21.

13. S. Humphries, *Hooligans or Rebels: An Oral History of Working-class Childhood and Youth, 1889–1939* (Oxford, 1981), including a splendid photograph of a street game, pp. 202–5.

14. T. Delaney, *The Roots of Rugby League* (Otley, 1984), pp. 13–14.

15. D.D. Molyneux, 'The Development of Physical Recreation in the Birmingham District from 1871–1892' (Birmingham MA, 1957, unpublished), p. 29.

16. N.L. Tranter, 'Sport and the Industrial Revolution in Scotland, 1780–1840: the

evidence of the Statistical Accounts', paper presented to the History Department Staff Seminar, University of Stirling, Autumn 1985.

17. M. Anderson, *Family Structure in Nineteenth-Century Lancashire* (London, 1971), p. 37.
18. P.J. Waller, *Town, City and Nation: England 1850–1914* (London, 1983), p. 76.
19. Cited by S. Meacham, *A Life Apart: the English Working Class 1890–1914* (London, 1977), p. 48; Ch. 2 contains a useful discussion of 'neighbourhood and kin'.
20. Fishwick, op. cit., p. 34.
21. Molyneux, op. cit., Appendix A and B.
22. R. Rees, The Development of Physical Recreation in Liverpool during the nineteenth century (M.A. Liverpool, 1968, unpublished), p. 69.
23. Mason, op. cit., p. 31.
24. D. Smith and G. Williams, *Fields of Praise: The Official History of the Welsh Rugby Union* (Cardiff, 1981), pp. 12–13.
25. Dr N.L. Tranter in the History Department, University of Stirling, is carrying out a major investigation of sport in nineteenth-century Stirling and kindly provided a list of football teams.
26. E. Hobsbawm, *Worlds of Labour* (London, 1984), p. 202.
27. J.H. Muir, *Glasgow in 1901* (Glasgow, 1901), p. 193. I am grateful to Dr J.D. Young for this reference.
28. D. Brailsford, 'The Locations of Eighteenth Century Spectator Sport', *Proceedings of the conference 'Geographical Perspectives on Sport'*, University of Birmingham, 7 July 1983.
29. S. Shipley, 'Tom Causer of Bermondsey: A Boxer Hero of the 1890's', *History Workshop Journal* (Spring 1983), 30; this article provides an innovative example of the fusing of oral testimony with conventional historical sources.
30. Mason, op. cit., p. 27.
31. Molyneux, op. cit., Appendix B.
32. Muir, op. cit., p. 185.
33. A. Ward and I. Allister, *Barnsley: A Study in Football* (Barton-under-Needwood, 1981), p. 3.
34. J. Crump, 'Amusements of the People: the provision of recreation in Leicester, 1850–1914' (Ph D Warwick, 1985, unpublished), p. 392.
35. Christopher Andrew, *The Listener*, 10 June 1982.
36. P. Joyce, *Work, Society and Politics: the culture of the factory in late Victorian England* (London, 1980); on paternalism see S. Yeo, *Religion and Voluntary Association on Crisis* (Brighton, 1976), and A. Redfern, 'Crewe: Leisure in a Railway Town', in J.K. Walton and J. Walvin (eds.), *Leisure in Britain, 1780–1939* (Manchester, 1983).
37. E. Hobsbawm, 'Mass-Producing Traditions: Europe 1870–1914', in E. Hobsbawm and T. Ranger (eds.), *The Invention of Tradition* (Cambridge, 1983), p. 301.
38. Murray, op. cit., *passim*.
39. Fishwick, op. cit., p. 162.
40. Mason, op. cit., pp. 158–167, and Fishwick, op. cit., pp. 176–90; the debate about crowd behaviour is still under way.
41. D. Morris, *The Soccer Tribe* (London, 1981), p. 111.
42. B. Butler, *The Giant Killers* (London, 1982), p. 47.
43. T. Delaney, *The Roots of Rugby League* (Keighley, 1984), pp. 13–14.
44. Mason, op. cit., p. 257.

5

Catalyst of Change: John Guthrie Kerr and the Adaptation of an Indigenous Scottish Tradition

J.A. Mangan

I

The Snell Exhibitions[1] to Balliol College, Oxford, which are made available to approved scholars born and educated in Scotland from the revenue of the estate of an Ayrshire landowner, John Snell (1629–67), have had a felicitous effect on Scottish education.

They have produced two noted Scottish innovators, one sung, one unsung: Hely Hutchinson Almond and John Guthrie Kerr. Almond is well-known,[2] Kerr little known, even in his own country. Both, it seems, were inspired by the English and 'Oxbridge' education of their period and contrasted it favourably with the limitations of their Scottish schooling. The evidence concerning Almond's attitude is clear. Mid-nineteenth-century Oxford's adoption of the post-Renaissance educational ideal of *'l'uomo universale'* impressed him greatly. He claimed that rowing with the Balliol eight was of more benefit to him than all the academic prizes he won at Oxford and Glasgow combined, and his biographer, R.J. Mackenzie, reported that the Isis 'opened his mind to a new set of values. His love of the open air, his passion for health, his appreciation of manly endurance . . . were the gifts of the river'.[3] The evidence concerning Kerr is more circumstantial and is to be found in his philosophical position as a schoolmaster, in his practical transformation of a Glasgow school and in his many pronouncements during his long period as headmaster about the virtues of a 'liberal education'. It seems that he was more inspired by what he saw than by what he did at Balliol. Ill-health kept him from the river.[4]

In both cases, however, the style of life at the ancient universities provided commitment to a wider philosophy of education than they had experienced in their own country, and ultimately the inspiration both gained from an English higher education had a considerable influence, initially, on two Scottish secondary schools in the east and west of Scotland respectively, and subsequently on the Scottish secondary system as a whole.[5]

By 1890 when John Guthrie Kerr became a headmaster, Almond had already been headmaster of Loretto for 27 years. He had won fame as an educational idealist and practitioner and his small school in Musselburgh was famous throughout Britain and the Empire.[6] Kerr never achieved such prominence although his own achievements were impressive. His school achieved a position of importance on Clydeside, and he became President of the Educational Institute of Scotland, and an Honorary Doctor of Laws of Glasgow University.

Kerr was born in 1853 and educated at Dumbarton Academy. In his last year at school he was awarded the Browne (Ayrshire) Bursary, and he enrolled at Glasgow University in 1871. He was an outstanding student, winning prizes for Maths, Greek and Natural Philosophy.[7] In 1874 he was awarded a Snell Exhibition to Balliol, Oxford,[8] where he remained for four years. Like Almond before him, his experiences there transformed his view of education. On returning to Scotland he was first a teacher of Mathematics at Kilmarnock Academy (1876–77), then a lecturer in Mathematics and Physics at the Church of Scotland Teacher-Training Centre (1877–90) and in 1890, he was appointed headmaster of Allan Glen's School in the heart of Glasgow.[9]

II

Allan Glen's had been founded in 1853 by a Glasgow businessman of the same name. He wished to create a school 'giving a good practical education to and for boys preparing for trades or business'.[10] In later decades it got caught up in the general process of reforms that characterized Scottish education at secondary level from 1820 onwards. Scottish secondary education developed differently from that of England in the earlier part of the nineteenth century. As Anderson has written, 'Scotland . . . had a national system of secondary schools which was compact and homogeneous enough to be discussed as a whole'.[11] These were its burgh (town) schools. They were urban day-schools controlled by town councils which responded positively to the utilitarian demands of a middle-class clientele. They had evolved from traditional grammar schools with a classical emphasis through a process of purposeful reformation in the first half of the nineteenth century. In some respects Allan Glen's fell neatly into this category. It was an urban day-school responsible to local need and demand. By the mid-1870s, in keeping with other leading secondary schools, it had changed both its status and its title. After

1872 the School Boards set up by the Education Act (Scotland) 1872 were able to accept the transfer of burgh schools with an advanced curriculum under the name of 'high class public schools'. Glasgow transferred its schools to the School Board a year later.[12] However, by 1890 in one important respect it was significantly different. In the early 1870s the Science and Art Department at South Kensington had given its approval to the creation of 'organized science schools' and 'five of these had grown up in Scotland, perhaps the most important being Allan Glen's, Glasgow'.[13] A private Act of Parliament in 1878 had transformed it from a school for the industrial classes into 'a secondary school, in which the utmost is made of science as a general instrument of education'.[14] It was certainly strong in science education, offering general science instruction up to the age of 14, and providing, in addition, a further two years' specialized training.

In one aspect of its life, however, at the time of Kerr's appointment the school was very much in the tradition of Scottish education. The school day was for academic effort and nothing else. Extra-curricular activities were eschewed. One former pupil, recalling his schooldays between 1884 and 1889, observed: 'There was no physical training or sports organisation of any kind. A cube of wood about two inches, obtained from the workshop, served as a football and was kicked about in the small playground to the severe detriment of your boots.'[15]

In short, there was no place for games in the formal curriculum, nor was participation in these activities part of a teacher's extra-curricular responsibilities. In Scotland as a whole, this state of affairs was increasingly under attack in the second half of the nineteenth century as the influence of the English schools spread slowly – first to Scotland's own residential public schools and later to the upper-class private day-school.[16] As Anderson has written of this period, 'the internal life of the schools saw an increasing emphasis on *gemeinschaftlich* features, on corporate life and *esprit de corps* . . . The character was now to be cultivated as well as the intellect, personal influence of the teacher over pupil replaced formal relationships, and schools began to taken an official interest in recreation and sport'.[17] Change was the outcome of criticism. In mid-century the rebukes of the commissioners of both the Argyll Commission and the Schools' Inquiry Commission, regarding the lack of opportunities for recreation and sport, were sharp. Harvey and Sellars of the Argyll Commission gathered statistics on amenities in 55 burgh schools and inspected 43 of them. They were not greatly impressed. In most cases they found playgrounds small, dirty and lacking adequate sanitary arrangements. They reported of Falkirk Grammar and Parochial Schools for example: 'The playground common to the two schools is miserably small. There are no covered sheds for wet weather, and the outhouses were in a very disgusting condition, and were quite insufficient for the purpose of decency and cleanliness'.[18] They saw at first hand, and with strong disapproval, the outcome of this arrangement, on a visit to Arbroath burgh school: 'The playground is very small . . . and not in grass. The consequence is that boys are driven out to play on the streets. The day of our visit they had taken advantage of the interval (between one and two o'clock) to organise a schoolboy battle in the street with the scholars of an adjoining school, to the evident annoyance of the lieges'.[19]

The cause of this state of affairs was clear to investigators such as Fearnon, Harvey and Sellars. In Fearnon's view, the schools being day-schools, teachers were only concerned with boys in class,[20] while Harvey and Sellars remarked in 1868 that the 'want of playgrounds is connected, no doubt, with the fact of these being exclusively day-schools. The relation of the scholar to the school, to the teachers, to one another is altogether different from that of boarders. The schools are places for work and work alone'.[21] They recommended more opportunities for the pupils to play games along the lines of the English public school system and saw virtue in the manliness, self-reliance, *esprit de corps* and healthy demeanour thus created.[22]

Their views had little impact on Allan Glen's in the following 22 years. And, in truth, there seemed little incentive for reform. The school flourished. In October 1889 the roll was 580 with an average attendance of 97 per cent. By September the following year it had grown to 655.[23] Kerr took up his appointment four months later. Things then began to change noticeably and rapidly. He did not share the conservative inclinations of his predecessors regarding the character-training potential of organized games.[24] He held a more liberal view of education.[25]

In his consideration of sport and the American mentality between 1880 and 1910 Donald J Mrozek wrote: 'The constituent groups which favored sport did so out of need. In different ways, each found in sport a strategy for regeneration and renewal'.[26] Out of specific experience grew specific visions of regeneration. For some it was physical improvement, for others the spiritual effects of physical training and for yet others the indistinguishable virtues accruing to mind and body.

In the west of Scotland, about the same time, Kerr had a firm and distinctive strategy of regeneration and renewal for Allan Glen's which appeared to lean heavily on his English experience. He was concerned to widen the boundaries of educational experience for his pupils, to include not merely the scientific and mathematical but also the physical. In his own words, delivered at his inaugural address as

PPP—D*

President of the Educational Institute of Scotland in 1894, he welcomed the recent 'violent changes for the better', which included physical exercise and 'a broader conception of the purpose of school studies'.[27]

It is not clear how far his innovative emphasis was determined by a concern with control.[28] There is no evidence of lack of control, no evidence of unruly behaviour, no record of the deliberate fostering of games as a means of producing exhausted, amenable pupils. The purpose, as far as the evidence suggests, was imitative, moral, and sparto-Christian, an uncomplicated adherence to the old cliché *mens sana in corpore sano*. Kerr was conscious of something missing in the Scottish education of his time – something valuable, and enjoyable. It appears to be as straightforward as that.

'Muscular Christian' is an overused expression but it is none the less an accurate description of Kerr. His endeavours to change the ethos of one dour and drab city school were remarkably similar, in some instances, to those other pre-eminent Kingsleyian stalwarts, Thring of Uppingham and Almond of Loretto. Like Thring he went out searching for suitable physical amenities for his boys. He found them to the south on the farm of Mr Alistair Young, Greenlees, Cambuslang, some miles from the city centre. And like Almond, he was not adverse to interrupting classes when keenness showed signs of flagging, and carrying off the pupils to the gymnasium to run and jump and hurdle, leading the way himself.[29]

His innovatory policy of *mens sana in corpore sano* was cultivated systematically[30] but it was part of wider 'civilizing' processes which involved a school song, concerts, magazines and more symbolic gestures, such as the creation of new school colours of light and dark blue combining quite deliberately the colours of two great English seats of learning[31] to which the pupils would now go in increasing numbers.

III

It is certainly true that in advocating and instituting change Kerr was part of a general movement throughout Scotland, briefly referred to earlier, which had as its aims the broadening of the experience of education among middle-class children – a broadening process incidentally inexplicably neglected by modern scholars. From 1860 onwards there was clearly an attempt to move closer to an 'English' concept of education prevalent there in both the public school and reformed grammar school,[32] and to establish a British model of secondary education, to the advantage of Scots seeking positions in the south. In some places such as Stirling High School, under the rector Andrew Hutchinson (1866–96), this process had been going on for several decades before Kerr changed things at Allan Glen's. But Hutchinson, it would seem, was less convinced of the importance of games in the curriculum than Kerr.[33] Kerr was closer in attitude to the 'heartier types with English backgrounds' whom Scottish school boards and governing bodies began to appoint in the 1890s and who made Scottish middle-class secondary education indistinguishable from its English counterpart. Men such as Edwin Temple (Glasgow Academy), H.J. Simpson (Glasgow High School) and Bingham Turner (Kelvinside Academy)[34] proved similar in inclination and enthusiasm to Kerr.

Most importantly, his professional predilections were clearly shared by many of his staff who loyally served the cause of school athletics and assisted in the transformation of education theory and practice at Allan Glen's in the late nineteenth and early twentieth centuries. Scots, educated in the traditional manner and without the benefits of a Balliol education, moved with the times, and assisted in the anglicization of their school and system. The reputation and standing of Allan Glen's ensured a stream of promotions to senior positions in other schools,[35] and in this way Kerr's values were disseminated throughout the secondary schools of Glasgow and the south.[36] An outstanding example of this process of diffusion is J. Jeffrey, an able athlete, who after providing 'strenuous aid' to Kerr in his physical ambitions for Allan Glen's, became rector of Selkirk High School and at both schools strongly endorsed the policy of physical training as 'an essential part of school education'.[37]

Of course, the extent of the process of anglicization should not be exaggerated. A distinct Scottish, and Glaswegian, identity was maintained by virtue of accent if nothing else. At the same time, as in England undeniably, reform occurred and 'unified the different elements of the middle class – around a professional rather than entrepreneurial ideal – marked the boundaries between its upper and lower strata, and absorbed families who could use new wealth to buy a superior schooling (which) led on to both English and Scottish universities, and to British careers'.[38]

Kerr himself was proud of the special, utilitarian emphasis of his Scottish science school. He had no foolish aspirations to model Allan Glen's wholly on the games-playing, classics-orientated schools of the English upper classes. He wished to provide the best elements as he saw them, of English and Scottish secondary education. In a cramped urban institution, so far removed in amenities from the more famous English public schools, his ambitions in some respects were certainly no different from those of his distinguished contemporaries Warre and Welldon who were located comfortably beside the Thames and confidently astride the famous hill. 'The strengthening of character, the securing of a sound foundation of general culture and the encouraging of a strong school spirit were ever before me', he once remarked, 'as the primary and essential objects.'[39] This originality, in a Glasgow urban context, was to be recognized and applauded in time. In 1910 for example, when Sir Henry Craik presented the address on Speech Day, he remarked that the school had been a pioneer in Scottish education, and Kerr as one of the leaders of the system had seen clearly that science was only one branch of education, and that you must train all the faculties.[40] Yet Kerr's aim was adaptation, not rejection. The special nature of Allan Glen's in the late nineteenth and early twentieth centuries was well described by Professor George Ramsay in a speech to the school in December 1910:

> The Chairman spoke of Allan Glen's as if it were a Technical School; that is not quite the right name to give it. It is not so much a Technical School as a Science School. It does not profess to teach the technical details of any special calling or profession – these can only be learnt in actual life inside the profession itself – what it professes to do is to use Science as an instrument for training and developing the mind combining with science just as much literary teaching and no more, as it is necessary for every educated man to possess. For this purpose science constitutes the main work, the foundation work of the school; it is pursued more systematically, more continuouusly, and in more subjects than in an ordinary Secondary School, the object being to give the mind a training through scientific methods and scientific reasoning as thorough, though different in kind, as that given through literary training in schools of the ordinary secondary type.[41]

In Kerr's obituary in the *Glasgow Herald* in 1932 much was made of his idiosyncratic interpretation of a liberal education – it was especially apposite to the industrial community of Glasgow. In the curriculum of Allan Glen's, we are informed, 'mathematical and practical science are relied on to develop and fortify the mind . . . Latin plays an unimportant part in the training and Greek, by regulations of the school, is expressly excluded'.[42] Kerr was always to insist that training in science and workshop exercises offered by the specialized curriculum of the school, was valuable, not so much on account of its direct utility to the boy in his career but rather because of its value as an aid to intellectual growth. This stance, the obituarist remarked, was a singularly effective reply to the 'classical' school.[43] To the very end of his life, long after he had given up the headship of Allan Glen's, Kerr maintained his belief in the pre-eminence of science work in schools. In a letter on the occasion of the Allan Glen's Old Boys Club Dinner in 1932 he declared: 'I stand by the Ramsay motto "Cum Scientia Humanitas" with Scientia as the inceptive agent. Further, I would add that intensive and specialized work has high cultural value in the fullest sense'.[44]

The educational act of union between Scotland and England was symbolically celebrated at Allan Glen's in the 1890s with the introduction of a school song to strengthen *esprit de corps*, which equally symbolically incorporated an odd, cacophonous line making honourable mention of indigenous tradition: 'we chip, file and turn very hard'.[45] 'Cum Scientia Humanitas' was now the motto for 'the new school'. In truth there was a mixture of idealism and realism in all this. Like English 'grammar school' headmasters of the time, Kerr was a willing agent of hegemonic infiltration.[46] He wished for elements of social style as well as ingredients of occupational utilitarianism. He dangled carrots for socially aspiring parents in the new suburbs of Bishopbriggs and Cambuslang, and the parents swallowed them. From 1890 onwards, the new image increasingly produced a new suburban clientele.[47]

Allan Glen's now achieved an harmonious balance between the forces of vocationalism and liberalism, which might well have served as model for the British public school and obviated the subsequent and bitter recriminations of its twentieth-century critics.[48] The school embraced the public-school games ethic and at the same time retained its traditional technical bias. The establishment figures who graced its Speech Day platform were not the archbishops, bishops, generals and governor-generals of dominion and colony who were to be found in large numbers at equivalent public-school ceremonies. The Certificates of the Science and Leaving Certificate Examinations of 1903, for example, were distributed by John Ward, President of the Institution of Engineers and Shipbuilders. Before Ward delivered the customary Guest of Honour's exhortation to the pupils, Dr Kerr was careful to stress his own particular ambitions:

> Attention had been given during the course of the year to the cultivation of athletics. As yet physical training could not be said to take a very marked place as an essential part of the curriculum but through the voluntary effort of some members of the teaching staff, it had been possible to develop a healthy attitude

to physical exercise, and the success of the school in inter-scholastic competitions, in general athletics, and in football, had been very gratifying.[49]

To redress the balance, for the conservative and 'utilitarian', he then summed up the list of recent academic successes at Glasgow and London. They make interesting reading:

At Glasgow University

B.D.	Alexander Pender Crichton, M.A.
M.D.	David McKail, M.B., Ch.B.
M.A.	James Reid, 1st class Honours in History
M.A.	Patrick Brough
B.Sc.	Pure Science, Hugh G. Robertson
B.Sc.	Pure Science, Andrew Edgar Struthers
B.Sc.	Engineering, Joseph Y. Walls
M.B., Ch.B.	Thomas Archibald
M.B., Ch.B.	John Cruickshank
M.B., Ch.B.	Walter Dawson
M.B., Ch.B.	Kenneth C. Middlemass
M.B., Ch.B.	Allan F. Miller
M.B., Ch.B.	Ralph M.F. Picken, B.Sc.
M.B., Ch.B.	Hugo G. Robertson

John Cruickshank was first in Honours and had distinguished himself by winning the Brunton Memorial Prize, this being the second time a former pupil of the school had won this prize. Ralph Picken was third in Honours, and Hugo G. Robertson was on the Commendation List, with distinction in Surgery and Materia Medica.

At London University

B.Sc.	John F. Mitchell, 1st class Honours in Physics
B.Sc.	Frederick Reid, 1st class Honours in Physics
B.Sc.	George A. Smiley, Honours in Chemistry

Besides these, James Bruce, presently lecturer in Chemistry at Huddersfield Technical School, had received his Doctorate of Science, and Dr Kenneth Duncan had gained his F.R.C.S. (Eng.). Wm. Bennett and James R. Thomson had won Lloyd's Register Scholarships at Glasgow University, and David Menzies had been appointed an Inspector of Factories. Perhaps the most interesting success on the list was that of George Gray, winner of the King's Prize at Bisley in July 1908, who, five or six years ago was one of the school's most distinguished athletes.[50]

This catalogue of achievement set the tone for Ward's address. His words, which reflected the concerns of perceptive and anxious monitors of educational fashion in the state school system, would have been singularly idiosyncratic if uttered before a public-school audience – no explicit reference to empire, no vaunting of the value of the games-ethic, no mention of the ideal of sacrificial service and no militaristic undertones. Instead, the pupils of Allan Glen's were given a sober synopsis of the historical role of the industrial professions and the future role of his young audience in maintaining Britain's industrial supremacy:

> In speaking to my fellow members of the Institution of Engineers and Shipbuilders last week, I tried in my Presidential Address, to trace some of the national benefits during the past century resulting from the work done by members of these professions. We who belong to the engineering or allied professions feel proud in having been pioneers in much good work; but we must recollect that there is no standing still among the nations, and the race for industrial supremacy goes on with accelerating speed. Our rivals represent many nations, and their efforts to outstrip us in technical and industrial science are steady and continuous. Our hope for maintaining our supreme position lies largely in the fact that young men, like yourselves, in every town and city, are being specially trained to efficiently master their professions, both scientifically and practically, and so rendered capable of grappling with, and probably solving, some of the problems which at present retard scientific and professional advancement.[51]

He was quick to assert, perhaps with a sideways glance at Kerr, the traditional values of Allan Glen's and its special characteristics – 'a sound and extensive training in science, in laboratory work, and in work-shop practice'.[52] This done, he was content to add generously that such a tradition was not inconsistent with a liberal education! He then sat down. Priorities were established, innovation was put in perspective! On Speech Day the following year, Mr James Mackenzie, Dean of the Faculty of Procurators and Convenor of the School Committee, treated the school in his opening remarks to a brief history of steam power for industrial purposes on the Clyde.[53] And in 1911, once again as chairman for the occasion, he provided an enthusiastic review of James Napier's *The Life of Robert Napier, Shipbuilder and Engineer*, founder of the Nairn Engineering and Shipbuilding Company.[54]

In the 'second city of the Empire' fittingly imperial careers were followed with pride but adjutants in Indian regiments and district administrators in Africa were in short supply. Instead the boys heard of the sterling efforts of former pupils in the Indian Public Works Board and the Colonial Medical Service. The school magazine made much of visits to the Electrical Engineering works in Glasgow, the Physical and Chemical Research Laboratory at Parkhead Forge and an 'expedition' by car to Blochairn Steel Works, Garngad, which included the following description – which contrasted oddly with the reports of former pupils pig-sticking, polo-playing or pacifying peasants on the frontiers of empire, to be found in the magazines of the public schools:

> The first item of interest was the row of furnaces in which the pig iron is heated along with haematite (Fe_2O_3). With the aid of blue spectacles we were enabled to see into one of these furnaces. We saw the molten mass in a state of brisk ebullition. The gas evolved was carbon monoxide, formed by the union of the carbon in the pig iron with the oxygen of the ore. After some time these furnaces are tapped from the back and the molten metal runs down a scoop into an immense ladle, which moves on rails. When the ladle is filled it is run along above a series of moulds, and these moulds are in turn filled by removing a plug from the bottom. When a sufficient amount of metal has been run into a mould, the flow is stopped by means of a lever, and the ladle moved along to another mould. The iron when it solidifies is in the form of ingots, weighing from eight to ten tons, which stand about 5 feet high and have a section of something like 1 ft. by 3 ft.[55]

Nevertheless, within the framework of traditional values, Kerr relentlessly and successfully proceeded to introduce southern fashion. A balance between science, arts and games was attempted. Early sessions of the School Society, established in 1892, contained essays and debates on 'America as a Manufacturing Rival', 'Science in our Century' but also 'British Orators' and 'Historical Fiction'. In the *Allan Glen's Monthly* for April 1911, juxtaposed proudly on opposite pages were a list of distinguished metallurgists from the school including Professor Andrew McWilliam of Sheffield University and an account of Ralph Erskine's recent success as featherweight boxing champion of the British public schools.[56] Kerr was a commonsense innovator. Throughout his tenure as headmaster, he was careful to ensure that technical academic awards were not adversely affected in the pursuit of liberalism. Standards were maintained. The School Prospectus for 1911 announced that during the previous six years,

Scottish competitors had won eight National Scholarships at the Imperial College of Science, all of them from Allan Glen's, and that during the previous 11 years former pupils had gained 160 university degrees (64 with honours or distinction), 151 prizes in engineering and 154 first-class certificates in medicine.[57]

Yet if technological expertise was admired and emphasized, the rhetoric of 'manliness' was promoted with equal insistence with the result that, in time, the pages of the school magazine bore a close resemblance on occasion, to those of the public schools of the south:

Play Up, School

When the teams are standing ready
   And the ball is on the ground,
Every player at attention
   As he waits the whistle's sound –
In the hush before the tumult,
   Ere there's time to break a rule,
Comes a murmur, quickly swelling –
   'Play up, School.'
Play up School, School, School,
Every man's to play the game and not the fool;
He's a member of the team,
And he's bound to 'put on steam'
When he hears the rousing shout of
'Play up, School.'

Kerr was not to be denied. As the *Glasgow Herald* on his death did not fail to make clear: 'No mention of R. Kerr would be complete that did not take account of his interest in athletics and his earnest conviction of the importance of the physical side of education. He diligently fostered among his boys a love for games and outdoor exercises'.[59]

IV

In his pursuit of English convention, Kerr was greatly assisted by the enthusiasm and support of George Gilbert Ramsay (1839–1921), Professor of Humanity in the University of Glasgow from 1863–1906 and a long-serving Governor of Allan Glen's.[60] Ramsay was the third son of Sir George Ramsay, ninth Baronet of Banff, and a product of Rugby and Oxford. All his life he was an enthusiastic athlete. *Who's Who in Glasgow* reported his interest in athletics of many kinds . . . golf, shooting, skating, angling, mountaineering and the like'. He won a reputation in the University as a muscular, athletic and energetic professor: 'Perhaps more skilful at cutting figures-of-eight than at the composition of Latin Verse'.[61] It was recorded in 1907 that with regard to Allan Glen's:

> From the first he has been the advocate of strenuous and carefully directed efforts in the gymnasium and the field of sport, to encourage whatever is manly in personal conduct and to teach the youth to carry himself like a future freeman of a great republic; these have been the ruling impulses of his benignant relation to the school. Arduous in study, arduous in sport, the motto might well be writ by his own favourite Horace. 'Rectique cultus pectora roborant'.[62]

In 1912, on the occasion of the presentation of a Ramsay Trophy for the Annual Sports, he announced that:

> He had thought . . . that he would like to do something to encourage the whole school to feel that the athletics of the school were an important matter, that the whole life of the school could be greatly enhanced and made something better if every one of them would strive, not only after individual distinction, but to do his best to raise the athletic tone of the school.[63]

And again the following year:

> To keep the body right was one of the essential duties of life: it was the aim of athletics to help the body to fit the mind to do its work. And there was something more than mere bodily care; there was a care for purity. The man who kept his body pure to keep his character pure, and the man who joined honestly and frankly in the exercises of a school, and who played his game fairly, had done much to secure for himself purity of mind as well as body![64]

Kerr himself wrote of Ramsay's contribution to the evolution of Allan Glen's in a letter in the school magazine on the occasion of Ramsay's death in the summer of 1921: 'He put before us the manly life, strenuous and cheerful, upright and honourable, unselfish yet self-respecting and answering at once to the call of duty'.[65] Ramsay, Kerr continued, had a great effect on his way of looking at life. His influence was substantial: 'Of those whose precepts and example have affected my life and my way of looking at life, there is no one of whom I think with more reverence and gratitude.'

Ramsay, for his part, never let an opportunity pass in which to lay stress on the all-round aspirations of Allan Glen's educational philosophy. On Speech Days his exhortations were peppered with claims for 'Athletics'. In his presentation of the school prizes in

7. George Gilbert Ramsay, Professor of Humanity, Glasgow University, and a staunch supporter of John Guthrie Kerr (*Allan Glen's Monthly*, Vol. XVI, March 1907)

8. Ralph Erskine, winner of the 1911 featherweight Public Schools' Boxing Championship and a pupil at Allan Glen's (*Allan Glen's Monthly*, Vol. XX, June 1911)

5. A pupil's suggestion for a new school badge, 1912, clearly shows the influence of Kerr's regime (*Allan Glen's Monthly*, Vol. XXI, June 1912)

6. Allan Glen's inter-scholastic athletes, 1913. John Guthrie Kerr is on the extreme left of the back row (*Allan Glen's Monthly*, Vol. XXII, June 1913)

January 1911, for example, he declared: 'It is difficult in the centre of a city like Glasgow to carry on athletic games in an organized and systematic manner, but every educationist knows what an effect properly conducted games have on developing manhood, courage, public spirit, respect for fair play, and acceptance of the principle of give and take, which have to do with the formation of character'.[66]

Kerr found supporters too in other even more influential quarters. 'The school's concern with character' won praise too from the Scottish Education Department. In its report for 1912, it observed:

> The school now conducted by the School Board of Glasgow under its old name of Allan Glen's School, and as the Glasgow High School for Science, continues to play an important role among the education institutions of the West of Scotland. In the ordinary subject of instruction, in athletics, and in the training of character, the interests of the pupils are well cared for.[67]

Kerr's most erudite and spirited defence of the new ethos is to be found in an edition of the *Allan Glen's Monthly* for June 1911, entitled simply 'Athletics'. It was both a celebration of the recent highly successful annual sports held at Hampden Park which attracted well over 4,000 'friends' and drew over 500 separate entries for the 41 events, and a forceful apologia for school policy: 'exercises of the body are essential to a complete school curriculum'. Rousseau, Gutmuths, Lagrange, Lucian, Homer, Lycurgus and Solon were marshalled as supporters of an 'athletic curriculum'. The plain verdict of history pointed 'to the inclusion of athletic training in the work of a school'.[68] The frontispiece of the same number was a picture of Ralph Erskine: 'member of Class 6 (Engineers); fine spirited, ½ miler, hurdler, high jumper, swimmer and gymnast; Vice-Captain of the school football team; and World's amateur 9-stone Boxing Champion' who had recently won the featherweight Public Schools' Boxing Championship at Aldershot. This famous victory over the English public schools was seen by Ramsay as the wholly logical outcome of the pursuit of a sound educational approach. 'I cannot tell you', he wrote to Kerr, 'how pleased I am to see the good school in an important contest taking a first place among the Great Public Schools of England . . . I congratulate you with all my heart knowing how you have cared for the physical as well as the intellectual side'.[69]

The *Glasgow Herald* was no less delighted with the victory of an obscure (to those in the South) Scottish school over the great schools of England:

> For many years boxing has been cultivated as a discipline in the great English Public Schools. The first of the schools

championships, instituted in 1885, fell to Cheltenham. From 1886 to 1890 Rugby and Haileybury divided the honours. During the past twenty years the championships have been won mainly by Clifton, Harrow, Charterhouse, Bedford, Dulwich, and St Paul's, the last-named famous four-century-old school having a full score of victories to its credit. On Friday, at the headquarters gymnasium at Aldershot, the public schools championships for 1911 were decided before a large gathering of spectators, including the Chaplain-General, Bishop Taylor Smith, and the Commander-in-Chief, Sir Horace Smith-Dorrien. The perform- ances all round were excellent, but the feature of the concluding day of the championships was the victory of Ralph Erskine, of Allan Glen's School, Glasgow, in the feather weights (9 st.). This school is on the list of endowed secondary schools, and has been recognised for some time by the Army Council as giving a training suitable for those who propose joining the Royal Military College.[70]

V

By the time of the Great War, Kerr's early ambitions were commonplace institutional purposes. They had become part and parcel of the unexceptional pursuit of 'manliness' as an end of school education, and were also the objectives of the wider Glasgow secondary school system. In December 1913, Dr Henry Dyer, Vice Chairman of the Glasgow School Board, gave the annual Speech Day address and urged 'true manliness' on the Allan Glen's pupils, finding a special few words to say about the mental, moral and physiological virtues of physical training in the school curriculum. Allan Glen's, he remarked, was not simply a specialist science school, it was building up physique and was now almost as distinguished for its athletic as its science training. Athletics, of course, he added, were not an end in themselves but a means to a greater end – 'the healthy body as the residence for the healthy mind'.[71] His views were echoed on the same occasion by the Reverend John G. Duncan, Vice-Convener of the High Schools Committee of the Glasgow Board of Education, who also had 'no doubt of the physical and moral benefit that could come from well-ordered physical training and from games carried on with enthusiasm and in a fine spirit',[72] while the Reverend J. Fraser Graham, Convener of the Committee on the High Schools of the School Board of Glasgow, and Chairman for the ceremony, remarked in a succinct assessment of the role of the school under Kerr, which must have been highly pleasing to him: 'While the work of the school had a distinctively scientific character, the great objective always present to the minds of those in charge was the laying of broad and sure foundations for liberal culture (and) the building up of energy and of character'.[73]

In 1917 Kerr resigned as headmaster of Allan Glen's to take up military nursing of the victims of the Great War.[74] He remained in close touch with the school, which throughout the 1920s and 1930s gradually adopted the athletic facilities, trophies, teams and record inseparable from a top-drawer school of the period. On Friday, 9 October 1931, the Kerr Memorial Plaque presented by the Old Boys' Club was unveiled at the school by a former pupil, now Principal of St Andrew's University, Sir James Irvine. Kerr replied to the many tributes he received with a 'stirring address' to the boys of the school: 'No obstacle – no progress; no fight, no joy of conflict' which included a valedictory summary of his long-held educational philosophy: 'It is in personal effort and a sense of contest, in the things of the mind or the muscles, that youth will best attain to intellectual manlihood, moral courage and physical fitness, and so promote efficiency for service to the world.'[75]

Kerr, no less than Almond, was a radical reformer and idealist. In contrast to Almond, however, he chose to work with indigenous Scottish educational tradition, to hold a balance between two

9. Sports Day under Kerr became a well publicized social event held at Hampden Park famous football ground (*Allan Glen's Monthly*, Vol. XXI, June 1912)

educational systems, to attempt to create a form of liberal education which assimilated a well-established scientific content. He broadened the curriculum of the school in the interests of health and 'character', and led a revolution in educational practice in the state schools of Glasgow which has still to be fully explored and recorded.

But more than this, Kerr's ambition for his pupils certainly included the intended erosion of traditional Scottish educational attitudes and it was, in part, a deliberate deracination process. He wanted to produce not simply recruits for Clydeside industry but Scots who would feel comfortable among English educated competitors and colleagues in the professions. To this extent his policy had, as a primary objective, the Anglicization and de-parochialism of his pupils. Adoption of the fashion of the playing fields was one significant means of achieving this.

In the history of the Scottish day-school he must rank close to Mackenzie of Edinburgh Academy as a significant agent of dissemination and change. He brought a philosophy and a practice to secondary education in Glasgow which consequently became the norm but which, in his time, were highly original. By means of his enthusiasm and energy he introduced English upper-class boarding school educational, ethical and cultural values into a west of Scotland middle-class day-school. He extended the definition, nature and scope of secondary schooling and, in doing so, won the approval and support of his professional colleagues. In any record of the history of secondary schooling in the west of Scotland on the basis of what is known, he is not to be overlooked, while the full extent of his influence on schools, society and culture in the area, both through his personal influence and the proselytizing zeal of staff promoted and dispersed through the region, has yet to be determined.

NOTES

1. See W. Innes Addison, *The Snell Exhibitions: Founder, Foundation and Foundationers* (Glasgow, 1901), pp. 1–28.
2. He has received considerable attention, see for example, R.J. MacKenzie, *Almond of Loretto* (London, 1905) and H.B. Tristram, *Loretto School; Past and Present* (London, 1910). Recently he has been discussed by Ian Thomson in his unpublished M.Sc. thesis, 'Almond of Loretto and the Development of Physical Education in Scotland during the Nineteenth Century', Edinburgh University, 1972, and in J.A. Mangan, *Athleticism in the Victorian and Edwardian Public School* (Cambridge, 1981). He has been further considered in J.A. Mangan, 'Almond of Loretto: Scottish Educational Visionary and Reformer', *Scottish Educational Review*, Vol. 11, No. 2 (November 1979), 97–106 and J.A. Mangan, 'Hely Hutchinson Almond: Iconoclast, Anglophile and Imperialist', *Scottish Journal of Physical Education*, Vol. 12, No. 3 (August, 1984), 38–40. A consideration of his attitude to imperialism is to be found in J.A. Mangan, *The Games*

Ethic and Imperialism: Aspects of the Diffusion of an Ideal (London, 1987), Chapter Two.
3. MacKenzie, *Almond of Loretto*, p. 16.
4. Letter to the author from the Archivist, Balliol College, dated 16 October 1983.
5. See Mangan, *Athleticism*, Chapter Three.
6. Ibid.
7. Addison, *Snell Exhibitions*, p. 22.
8. *Alumni Oxoniensis*, 1715–1886, p. 146.
9. *Roll of Graduates, University of Glasgow, 1727–1897*.
10. Joseph Rae, *History of Allan Glen's* (Glasgow, 1954), p. 25.
11. Robert Anderson, 'Secondary Schools and Scottish Society in the Nineteenth Century', unpublished paper delivered to the Scottish History of Education Society, Autumn, 1985, p. 3.
12. Education Authorities replaced School Boards in 1918. See James Scotland, *The History of Scottish Education*, Vol. II (London, 1969), pp. 37 and 62.
13. H.M. Knox, *Two Hundred and Fifty Years of Scottish Education 1696–1946* (Edinburgh, 1953), pp. 134–5.
14. Balfour Committee First Report, pp. 177–8, quoted in Robert Anderson, *Secondary Schools and Scottish Society* (Oxford, 1981), p. 194.
15. Rae, *Allan Glen's*, p. 48.
16. Anderson, 'Secondary Schools and Scottish Society', p. 20.
17. Ibid.
18. See *Argyll Commission*, Third Report, Vol. II, p. 160.
19. Ibid., p. 39.
20. A comment that illustrated a traditional Scottish approach to education: academic, earnest and joyless. To quote Thomson ('Almond of Loretto . . .', p. 26), 'the curriculum left little room for play or frivolity'.
21. *Schools Inquiry Commission*, (Scotland) Vol. VI, 1, 1868, p. 40, quoted in Thomson, 'Almond of Loretto . . .', p. 57.
22. *Schools Inquiry Commission*, (Scotland), Gerard Report, Chapter III, p. 52.
23. Rae, *Allan Glen's*, p. 51.
24. Ibid., p. 133.
25. Ibid.
26. Donald I. Mrozek, *Sport and American Mentality 1880–1910* (Knoxville, 1984), pp. 3–4.
27. See the Presidential Address in the Report of the Proceedings of the E.I.S. at the 49th Annual General Meeting held in the High School of Edinburgh, Saturday, 21 September 1895, in *Reports of the Educational Institute of Scotland 1892–96*, Scottish Record Office, pp. 90, 342, 119.
28. For a discussion of the relationship between games and social control, see Mangan, *Athleticism*, Chapter Two.
29. Rae, *Allan Glen's*, p. 73.
30. Ibid.
31. Rae, *Allan Glen's*, p. 59.
32. Mangan, *Athleticism, passim*, and 'Imitating their Betters and Disassociating from their Inferiors: Grammar Schools and the Games Ethic', paper delivered at the *Annual Conference of the History of Education Society*, December, 1983, *passim*.
33. Anderson, *Secondary Schools and Scottish Society*, p. 23.
34. Ibid., p. 24.
35. See *Allan Glen's Monthly*, Vol. XVIII (January 1909), p. 1, and the reference to promotions by M.G. Ross, Convenor of the School Committee at Speech Day, 23 December 1908.
36. *Allan Glen's Magazine*, Vol. XX (February, 1914), p. 4.
37. *Allan Glen's Monthly*, Vol. XVI (1906), p. 6.
38. Anderson, *Secondary Schools and Scottish Society*, p. 27.
39. Rae, *Allan Glen's*, p. 46.
40. *Allan Glen's Monthly*, Vol. XIX (January 1910), p. 3.

41. *Allan Glen's Monthly*, Vol. XX, (January 1911), p. 2.
42. *Glasgow Herald*, 30 October, 1937, p. 13.
43. Ibid.
44. Rae, *Allan Glen's*, p. 106.
45. Ibid., p. 55.
46. See Mangan, 'Imitating Their Betters . . . .', *passim*.
47. Under Kerr and later headmasters the school increasingly catered for the Glaswegian middle classes of the outer suburbs. Fittingly the school's playing fields moved to the then semi-rural suburb of Bishopbriggs after the Great War.
48. See, for example, Corelli Barnett, *The Collapse of British Power*, (London, 1972).
49. *Allan Glen's Monthly*, Vol. XVIII (January 1909), p. 2.
50. Ibid.
51. *Allan Glen's Monthly*, Vol. XVII (January 1909), pp. 2–3.
52. Ibid.
53. *Allan Glen's Monthly*, Vol. XX (January 1911), pp. 1–3.
54. *Allan Glen's Monthly*, Vol. XXI (April 1912), pp. 1–4.
55. *Allan Glen's Monthly*, Volume XX (April 1911), p. 7.
56. Ibid.
57. Ibid.
58. *Allan Glen's Magazine* (Spring 1921), p. 13. This was by no means an isolated example, see, for example, *Allan Glen's Magazine* (Midsummer 1922), p. 15, but especially *Allan Glen's Monthly* (June 1906) p. 7.
59. *Glasgow Herald*, 30 October 1937, p. 13.
60. It is rather an oddity that Ramsey, President and a leading inspiration in founding the Classical Association in Scotland to maintain the place of Classics in Scottish Education, should have been such a staunch supporter of the utilitarian Allan Glen's.
61. *Bailie*, 28 January 1874, p. 15.
62. *Allan Glen's Monthly*, Vol. XVI (March 1907), p. 1.
63. *Allan Glen's Monthly*, Vol. XXI (April 1912), p. 10.
64. *Allan Glen's Magazine* (Spring 1913), p. 11.
65. *Allan Glen's Magazine* (Midsummer 1921), p. 3.
66. *Allan Glen's Monthly*, Vol. XX (January 1911), p. 5.
67. *Allan Glen's Monthly*, Vol. XXI (1912), pp. 2–3.
68. *Allan Glen's Monthly*, Vol. XX (June 1911), pp. 1–3.
69. Ibid., p. 2.
70. *Glasgow Herald*, 30 October 1937, p. 13.
71. *Allan Glen's Magazine*, Vol. XXIII (February 1914), p. 3.
72. Ibid., p. 4.
73. Ibid., p. 1.
74. He moved to Lancashire in 1917 to the post of superintendant of an institution for the rehabilitation of wounded soldiers, and died in 1937, aged 83, at his daughter's home in Blackburn.

6

Brothers of the Angle: Coarse Fishing and English Working-Class Culture, 1850–1914

John Lowerson

I

Historians of mass sports, leisure and popular culture in industrializing England have proved remarkably short-sighted in their treatment of the country's largest participant outdoor sport, angling; so short-sighted, indeed, that they have virtually failed to notice it at all.[1] This may be because the obvious capitalization of and moral problems inherent in games like football, of entertainments and the music hall, or the continuity of 'rough' pursuits have proved more amenable to the development of grand hypotheses than an activity characterized by its apparent individualism and passivity. This essay is intended to redress that balance and to demonstrate the essential inter-relationship of angling as a popular recreation with wider aspects of class identity, attitudes to work, sporting ethics and the use of disposable income in a mature industrial society. Above all, it was a product of wholesale urbanization and could only have developed with the railway age.

I

The present pre-eminence of angling as a leisure pursuit (some four million anglers in the 1980s) poses an immediate problem; it is virtually impossible to provide an exact chronology of growth because the only adequate index, the licensing of fishermen, was not applied to the vast majority until the 1960s.[2] Yet it was quite considerable early on. The *Fishing Gazette* estimated that there were 50,000 anglers in 1878, even conservative assessments suggest 100,000 by the early 1900s and 200,000 would not be too unreasonable by 1914, the closing date of this study.[3] Sheffield alone had over 20,000 in its affiliated clubs by that date, although, as we shall see, the ratio of participants to that town's population was exceptional.

Like the other major field sport inherited from pre-industrial society, fox hunting, fishing was pan-class in its following but, whereas fox hunting was essentially financed for its élitist values by an aspirant upper-middle class, angling relied on active participation by all social groups and divided the classes differently; both the quarry and fishing areas could be delineated in broad class terms.[4] With some exceptions noted below angling was not a sport in which either overt attempts at hegemony or class hostility were common; it did allow for a particular extension of working-class ambitions in uneasy parallels with prevailing middle-class attitudes during the later nineteenth century. Freshwater fishing offers a basic divide which has allowed wealth and class to determine broad bands of species availability, until recently, between 'game' fish (salmon, trout and grayling) and 'coarse' fish (virtually all others). The former were restricted, by the mid-Victorian period, to purer waters and surrounded with a mythology, generated by writings such as Charles Kingsley's *Chalk-stream Studies*, of demanding tremendous skills, fishing with artificial flies as distinct from the live baits (especially maggots) or food pastes used generally for their more despised relatives.[5]

The élite connotations game fishing accumulated led to a spectacular growth. In 1879, the first year of trout rod licensing, 9,109 permits were issued in England and Wales; by 1911 that figure had risen to 55,069.[6] Rising middle-class disposable incomes and disposable time combined with the pursuit of exclusive symbols and a powerful self-persuasion of the need to relieve urban stress to foster a demand for game-fishing waters that was increasingly difficult to satisfy. After a London bookseller, Henry Wix, founded the Amwell Magna Fishing Club in 1831 to preserve trout waters on Hertfordshire's River Lea, some 20 miles from the capital, pressure on good waters grew.[7] By 1900 a season's fishing on Hampshire's River Meon cost £50, and demand could still not be satisfied.[8] Enterprising commercial firms, such as the developers of the Bellagio Bungalow estate in Surrey and the Norton Fishery at Baldock, Hertfordshire, stocked local streams and ponds and charged anything from 12 to 35 guineas for a season ticket.[9] Elsewhere, small middle-class groups such as the Sheffield Derwent Fly-Fishing Club took over the purer waters near to industrial towns for weekday fishing. When these proved insufficient similar bodies persuaded the late Victorian and Edwardian public water authorities to stock their reservoirs and allow limited access; humbler anglers were excluded both by the costs and by the assumption that they would prove a health risk.[10]

Access problems such as this were occasionally exacerbated by overt class conflict. When the professional men and prosperous tradesmen of Whitby decided to preserve the River Esk for salmon and sea-trout fishing they clashed immediately with local miners who had enjoyed free fishing as a customary right. They compromised by keeping a small area for them, rescinding the concession in 1868 when it was discovered that the recipients were selling fish to augment their incomes.[11] In the 1880s Hull's Wilkinson Angling Association split when some members tried to exclude all working men from joining.[12] Other conflicts arose from class-ascribed patterns of angling behaviour, such as when W.B. Webster claimed in the *Field* and *Fishing Gazette* that he was driven from a peaceful afternoon's meditative pleasure fishing by 17 rowdy members of the Brothers Well Met Angling Society who were trying to organize a competition; they replied that his was a typical example of upper-class arrogance.[13] This clash between middle-class 'pleasure' anglers sharing a common perception of working-class fishermen as noisy and competitive was repeated often, despite there probably being a substantial number of working-class pleasure anglers too. The better-off coarse fisherman in search of relative solitude could usually find more remote waters and pursue large 'specimen' fish such as carp or pike, leaving the others to fish for the smaller, shoaling species. Middle-class pressures helped develop the Norfolk Broads as a multi-purpose water sports area and spurred along private local protective legislation for Norfolk and Suffolk in 1877.[14] Occasionally, the sense of class division proved ironic. When William Morris, the designer and socialist pioneer, went fishing during a visit to the Cobden family in 1889 he caught 27 gudgeon and seven 'fine perch', which were stolen; 'We agreed that, as a Socialist, he should have rejoiced that others had enjoyed his fish at their evening repast.' At which, he lost his temper![15]

That there was a broad class divide in freshwater fishing was widely perceived by many white-collar and manual workers. They were sometimes forced to find waters which could be 50 miles or more from their homes, and they occasionally articulated their resentment. The mass meetings which helped to support Mundella's close season legislation in 1877–78 were revived to prevent the imposition of rod licensing for coarse fishing in 1896. The one shilling fee proposed for the Yorkshire area aroused fury at large Sheffield meetings where the 'rod tax' was castigated as a product of game fishing legislators, 'those who toil not, neither do they spin. But perish the thought that the more democratic army of coarse-fishers should pursue their favourite recreative pastime unscathed.'[16] The proposal was dropped but it had illustrated only too clearly that social and spatial segregation of the majority of English fishermen which fostered competitive coarse angling.

II

Fishing as a sport developed, as R.C. Hoffmann has demonstrated, on the European mainland in the high middle ages but found its strongest roots in England from the sixteenth and seventeenth centuries.[17] Among poorer anglers there must always have been a symbiosis between sport and food and this continued well into the nineteenth century; in Victorian Nottingham roach and dace, virtually the smallest fish pursued, were sold in the local markets for 3d a pound.[18] Essentially a rural recreation, it had long enjoyed a refined literary support, represented for many by a classical text: Izaak Walton's *Compleat Angler*, first published in 1653. The book's circulation appears to have been socially limited, not least by cost, until a whole range of popular editions appeared in the later nineteenth century.[19] Walton offered practical advice, folklore, High Anglicanism and one of the most consistent statements of the persistent rural Arcadianism which has been variously described as one of England's greatest strengths and a major factor in its prolonged twentieth-century industrial decline. He became for many working-class fishermen, who probably never read him, an arcane symbol of an anti-urban idealism to whom frequent lip-service was paid by toasts at annual dinners and in the naming of such clubs as the Walworth Waltonians or the Isaac Walton Angling Society of Kingsland Road, London.[20] Around him grew myths which allowed for a widespread characterization of personality types as a buttress for the sport: when 400 people attended the Rother Anglers' Club ball in February 1893 it was claimed that: 'Anglers are, without doubt, pleasure-loving people. They inherit the faculty from genial Isaac . . . as appreciative of mirth and fun as a frollicking schoolboy.'[21]

Arcadianism caused problems when it affected the work-patterns of mature industrialism, but the development of popular angling fitted in very well with the notions of ethically acceptable working-class recreations chronicled by Bailey and Malcolmson: 'Angling has long been a pastime much practised by miners, furnacemen, and others engaged in industrial occupations in the Midlands, and a love for the contemplative man's recreation has done much to tone down the taste for more vicious forms of amusement.'[22]

In the 30 years after 1870 it could be claimed, not unreasonably, that the competitive institutionalization of this contemplation had turned the 'very scum of the Sheffield dregs' into paragons of order.[23] The problems with repeating such claims, widely distributed in late Victorian angling writing, is that they fit into simplistic models of a middle-class hegemony far too easily. It is more likely that the sport's rapid spread in the last quarter of the nineteenth century owed much more to a moderate shift in basic urban working-class cultural patterns, with an instrumental use of apparent organizational networks on *bourgeois* models, than to any passive shepherding into folds of recreational acceptability. There were some clubs sponsored by employers, such as railway companies in the London area or Sheffield steel firms such as J. Rodgers and Sons, but the vast majority grew out of much older roots, the pubs and their associated friendly societies.[24] Because of the sheer density of pub distribution in most working-class areas, clubs reinforced what Richard Holt has outlined as the essentially street-based *milieu* of most working-class sports; the very workplace/residential proximity of northern towns in particular meant that the work mate, job-centred foci of male status were carried over into the founding and running of clubs. Publicans welcomed them, happily hanging club name boards outside their houses, because the use of a separate room in the pub, usually lined with stuffed fish in glass cases, meant a solid and not too disruptive clientele.[25] In response to the Victorian temperance onslaught anglers were often keen to play down the drink element, while stressing and preserving conviviality. As one Sheffield cutler told a Parliamentary committee on the Mundella bill in 1878: 'There was one temperance society out of two hundred (laughter). They mostly met at public houses and had a jolification (*sic*) afterward, without exceeding the bounds of temperance'.[26]

The sport's growth led some morally proselytizing bodies, such as churches, to offer their own alternatives to a pub base, verbally rejecting yet essentially accepting the bonding patterns of local life. It is probable, however, that organizations such as Sheffield's Attercliffe Zion angling club or the Bradford Total Abstinence Anglers, remained a very uneasy minority.[27]

The local roots and the space limits of pubs, together with the wish for face-to-face management, kept angling clubs small, usually with 40 as an upper membership limit. Identification with pubs was common in club names – London's Stanley Anglers met at the Lord Stanley in Camden Park Road; in Leeds the Rockingham was founded at the pub of that name in 1879, with the landlord, E.F. Atkinson, as president and treasurer, and with a continuation of the older fund holding of the friendly societies.[28] The comradely sense of the latter carried over into a widespread use of 'Brothers' in club titles; W.G. Callcutt noted it as the most common element in London's 620 clubs, such as the Great Northern Brothers of Clerkenwell, the Bermondsey Brothers, or the Brothers Well Met whom we have already seen in bankside conflict.[29] This brotherhood was a common element in angling reportage, whether in the rules of the Manchester Amalgamated Anglers' Association: 'Far away from the noise and bustle of towns, the Angler enjoys himself, undisturbed by anything but the echoing of distant waters . . . . welcoming his brother Angler',[30] or in the value attached to annual dinners, offering rivalry without jealousy and 'brotherly relationship'. None the less, clubs could and did split in most unbrotherly fashions. In Manchester the Dutch Birds Angling Society had to threaten legal action to stop a breakaway group from annexing its waters and London's Warwick Angling Society came into being in 1900 after trouble in the Walthamstow Brothers' Angling Society.[31] True to type it called itself after a former lord of the manor, met in a pub named after his son (the Lord Brooke), adopted as its motto *Juncta Juvant* (United they assist one another) and aimed at 'cultivating social feeling'. When London's clubs finally united in a common association in the 1880s the sense of brotherhood was consciously reinforced by a system of area visiting, where clubs hosted each other on their weekday evenings, often in preparation for 'friendly' matches on the river.[32]

Although coarse angling could be welcomed as a desirable adjunct to the demands of industrial work-discipline, its development as an essentially working- and lower-middle-class sport organized by its participants raised questions as to how far fishermen were prepared to co–operate with those values. It actually reinforced opportunities for flouting prevalent middle-class practices or for a very selective process of their adoption. In the south, in particular, angling became clearly identified with working-class anti-sabbatarianism. I have examined elsewhere the process by which sport became a measure of middle-class apostasy in the later Victorian period;[33] in the present case it was not so much a case of following the bad example of their betters as of a reinforcement through bureaucratization of a rejection of organized religion which had become only too apparent in the 1851 religious census. In the London area the attraction of the Thames and its tributaries for all water sports on Sundays grew especially strong in view of the restrictions imposed on access by the six-day working week; even with the wider availability of Saturday half-day holidays after the 1870s the time demands of travel, and the need for adequate hours of daylight, limited their value for many workers. In 1860 the Friendly Anglers of Shoreditch broke away from older groups to become the first working men's club devoted to fishing on Sundays only and was soon followed by many others, not only on the Thames.[34] From the 1870s several hundred London anglers descended each Sunday in season by special train to fish near Amberley in Sussex, singularly visible in a still largely church–going countryside.[35] Inevitably, a sporadic sabbatarian war wandered through the pages of the angling press. While the Surrey Piscatorials and the Freedom Angling Club of Sheffield banned Sunday fishing and the Manchester Anglers' Association discountenanced it,[36] London provided a number of apologists, such as Francis Frances, for whom social utility and a simple pantheism ran hand-in-hand. Against 'airy nothings from the pulpit' were placed the value of sermons and prayers from stones and flowers, listened to by 'a poor man on a Sunday in a snug corner on a pollard, with his little family about him'.[37]

In the north these tensions became much more apparent because a popular but non-church attending sabbatarianism reinforced traditional leisure practices inimical to the total implementation of new industrial ideals. In 1896 the Sheffield Anglers' Association split after a reported threat by a temperance group to impose Sunday observance.[38] In fact, as we shall see, the issues were far more complex, but the confrontation demonstrated the uneasy relationship between official sabbatarianism and its popular and unofficial version. In Sheffield and its areas, the sacrosanct day for fishing until 1914 was Monday, and the rapid spread of clubs and their affiliation actually reinforced pre-industrial work patterns rather than, as in such sports as football, tailoring an inherited popular form to fit changing disciplinary modes. While Saint Monday declined sharply in much of England after the 1870s, it stayed healthy in Sheffield. 'Of course Monday is the great day of the week, Sheffielders being great patrons of Saint Monday', the *Fishing Gazette* claimed in 1878 and this still held largely true 30 years later, despite the fact that the town had partly recognized new circumstances in naming one of its principal football clubs, Sheffield Wednesday, after its commercial half-day.[39] In 1900 81 per cent of the fishing matches organized from the town took place on Mondays; thereafter there was a marginal decline, but the figure still stood at 69 per cent in 1912, the last year for which data are available.[40] Fishing paralleled other popular recreations in the town, where Monday was also devoted to competitive live pigeon-shooting in prize matches arranged by publicans for spectators with considerable side-betting.[41] The town's distinctive industrial structure, with its key workers employed in small cutlery and engineering workshops working at an older rhythm to self-imposed targets, made the imposition of any factory discipline difficult, and even many of the larger works appear to have had to accept a compromise on this issue. The paradox was further heightened by the fact that the peak months for competitive angling, July, August and September, were those that coincided with maximum demand in the metal trades.[42] Sheffield's Edwardian prosperity was built on a fine calculation by many workers of their income needs related to time worked, markedly at odds with any notion that wealth and production had to be determined by an absolute utilization of all available working hours. In a sense Wiener's thesis that arcadianism contributed to England's rapid decline as a mature economy holds true in this instance but, it would seem that in Sheffield an important sector of the work–force had never bothered to reach the position from which that decline could begin.[43]

III

The sport's development took on clear area differences. The growth of an infrastructure inevitably reiterated these and provided further illustrations of the local rivalries and regional tensions which appear irremovable from the English context, not least between north and south. At the simplest level there were distinctive regional patterns occasioned by the nature of the available waters. London clerks tended to fish the Thames west of the brackish tidal flow, the Cockney manual workers the River Lea in Essex and Hertfordshire, the south Londoners the waters of the Sussex Arun. The width of these and the commonest quarry, the diminutive roach, demanded long fishing poles and very light tackle. The Thames offered a 'democracy', easy of access to those who could afford the fares.[44] A similar ease was experienced in Nottingham where the standard rod was 11 feet long, casting relatively heavy tackle to cope with the fast-flowing River Trent, which traversed the town.[45] Around Sheffield the pure waters of Derbyshire were almost entirely annexed by a tiny minority of trout fishermen, reckoned at less than one per cent of the town's anglers, and the main river, the Don, was so polluted that fish were virtually extinct.[46] Working-class anglers, forced to find waters at least 50 miles away in Lincolnshire and often much further, developed a style suited to slow marshland and fenland rivers; they used 'short' rods, silk lines and quill floats, light tackle for the rapid catching of large numbers of small fish since most matches were judged by weight caught. The standard Sheffield bait consisted of handfuls of 'gentles' (maggots) coloured with bright dyes and pierced with fine wire hooks the manufacture of which by specialist needle firms such as Allcocks of Redditch is an area of business history too considerable for this essay.[47] When national matches began officially in 1906 the relative merits of these local styles proved an endless source of oral and printed debate, serving as a further bonding for fierce local patriotisms.

Sheffield was widely regarded as the key coarse fishing centre, in terms of numbers, style and organization – a singular irony in view of its spatial divorce from its waters. The first great burst of angling's popularity up to 1880 was reinforced by the *Fishing Gazette*'s adulation of virtually everything concerned with the town. By 1878 it was reckoned to have some 8000 anglers, compared with London's 10,000; in terms of proportions Sheffield outnumbered the capital eleven to one, with a ratio of one angler to every 37 of its population, as against London's one for every 400.[48] Although later writers could characterize many of the early Sheffielders as scum, a much happier picture was usually painted of the 'phlegmatic nature of the Sheffield artisan': 'Seated on the ever-present basket with rod in hand, he sits lazily watching the smoke that curls from his "cutty" [pipe], and through it keeps his eye fixed on his quill'.[49]

This idyll must be set, however, against other reports of Sheffielders' 'war whoops' every time a fish was caught.[50] Fishing was claimed to be responsible for the rapid late Victorian improvement in the town's mortality rates.[51] If that must be treated with caution, the sport was a clear item of conspicuous consumption for men in the metal trades where average wages were reckoned at 37 shillings a week (£96 a year) at the turn of the century, even when interrupted by Saint Monday.[52] A simple rod cost about 10 shillings, an outfit much more, and to this had to be added replacement tackle and travel costs. Despite its popularity in the town's pub world and its high visibility in the national angling press, the sport virtually escaped notice in middle-class writing on the town's virtues. Two major Edwardian studies managed to avoid it altogether and the area's recent historians have been just as blinkered.[53] Sheffield was generously provided with the standard paternalist amelioratives of grim industrialism, such as parks, art galleries and libraries, but the most significant attempt by the town's work–force to provide its own escape from being corralled into them was glossed over.

Sheffield was the first town to achieve some form of angling institutionalization similar to that emerging in team games and athletics from the 1860s, but without their tendency towards minute regulation. In 1869 the leading clubs formed the Sheffield Anglers' Association as a collaborative bargaining base for acquiring access to waters and travel concessions. Within a decade it had 8000 affiliated members in 180 clubs and was run from the Crown Inn in Scotland Street, where the 'anglers' House of Parliament' was surrounded by the pea-soup fogs and shoeless children of one of the town's poorest areas.[54] Clubs joined slowly over the next 20 years, reaching 223 by 1890, until the early years of the twentieth century brought a very sharp increase; from 13,700 affiliated members in 1901, and 17,646 in 1906–7, to 21,291, grouped in 500 clubs on the eve of the First World War. Its most publicized acts were to prompt and support Mundella in 1877–78; since he was a Sheffield MP, his introduction of a private bill for a close season drafted at his expense was surely a clear recognition of the local political situation after the 1867 Reform Act. In conventional terms, angling remained largely outside the political arena (there has been no attempt, if the evidence remained, to assess the sporting activities of contemporary working-class activists), but the 1878 experience and angling's revival to reject the 'rod tax' in 1896 do represent a significant appreciation of pressure group tactics.

The 1896 agitation occurred in the midst of an organizational crisis for the Association. After its initial spurt it was claimed to be less responsive to change than later organizations elsewhere, to be 'old fashioned' and 'behind the times', with a strong tendency towards a self-perpetuating oligarchy in the shadow of its secretary since 1869, T. Walker. Disagreement finally focused on the strong pub links, the drinking on club nights and a suspicion of a drift towards Sunday fishing. It was the pressure for respectability in the sport that led to demands for its quarterly meetings to move to the Temperance Hotel, away from the smoke-filled Crown Inn. A small group of reformers was voted on to the committee in 1895 but displaced at an angry general meeting the following year, when 'there was a good deal of feeling displayed' and, according to a local legend difficult to verify, police were called to restore order. The reformers' leader, H. Liddell, was defeated in the secretarial election by 135 votes to Walker's 253. Bitterly angry, he left to found a rival body, the Sheffield Amalgamated Angling Society.[56] It soon acquired 2000 members and still survives, although it never came close to rivalling the older body before 1914, if only because, despite loud disclaimers, it was popularly associated with temperance and denominational sabbatarianism. Within a decade it was sitting in relative amiability on regional bodies together with its rival; fishing appeared to operate in relative harmony when early flurries died down.

Sheffield's pioneering work was widely copied – Stockport had 1200 affiliated to its association by 1878.[57] London developed a number of area associations, usually at loggerheads with each other. A federation, the London Anglers' Association, was virtually forced on them by the railway companies in 1884, eventually bonding 620 clubs together.[58] It avoided a decorative superstructure and was run by 'working anglers' such as its president, Philip Geen. Like the other major associations, it limited its interference in the running of local clubs but it did possess some disciplinary powers. Apart from match regulation and the approval of an annual timetable of competitions, these bodies could actually determine a club's survival by refusing access to protected waters and the purchase of concessionary railway tickets. The London Association forced the Clerkenwell Amateurs out of being when it decided the latter was not a *bona fide* club.[59] This sanction was rarely used but was a feature of all the associations, such as Manchester, Birmingham, Leeds and so on, that followed the Sheffield model.[60]

The drive to form urban associations was almost entirely instrumental, arising out of a realization of the value of collaborative bargaining powers rather than from any general notions of wider brotherhood as such. Two issues prompted this process – access to waters and the cost of transport. The sheer pressure of the sport's growing popularity by the 1870s forced working-class anglers to look for all available waters. In London the westward spread of suburbia and the expansion of Thameside towns like Maidenhead for bankside villas reduced fishing space and the Lea was also threatened. By 1879 one London body had already negotiated use of the Grand Junction canal at Brentford. The new association went on to acquire rights on the River Arun, the Huntingdon Ouse and waters around Ely in Cambridgeshire, 70 miles away.[61] Sheffielders used initially the few clear lower stretches of the Don and the Trent but spread gradually into Lincolnshire, fishing the Stainforth canal from Crowle to Keadby and the River Ancholme at Brigg. It was not initially a costly process since most coarse waters never acquired the financial significance for depressed agricultural landowners that the trout streams did. Access was cheap, often free, and carefully negotiated between deferential clubs and associations and landowners. Magnates such as Sir Walter Barttelot in Sussex were quite happy to allow Londoners access to their waters in the 1870s, although mass use and competition for picturesque landscapes in such middle-class retreats as nearby Amberley seem to have prompted a later retrenchment.[62] The *Fishing Gazette* recognized early on that this was probably the single most important role of many clubs for northerners, and membership was often the only way of obtaining access to the pleasure angling that more fortunate southerners had taken for granted before the spread of suburbia.[63] By 1879 Sheffield was moving southwards into Lincolnshire, renting four and a half miles of private waters on the Trent at Torksey, and negotiating to share 40 miles of water with the Lincoln Anglers' Association.[64]

It was not always easy; Sheffield's bitter history of trades union disturbances often made it difficult to arrange access for matches, and some worried landowners would demand a £30 deposit before allowing a single event.[65] In the Edwardian period Sheffield had rights on 120 miles of Lincolnshire waters and joined with the Lincoln, the Sheffield Amalgamated, the Leeds and 14 other associations to form the Witham Joint Angling Committee, leasing 40 miles for ten shillings a year.[66]

Not only landowners had to be won over – in the Fens one local club resisting Sheffield encroachment found itself at odds with local traders who welcomed the custom.[67] From acquiring rights, it was a short step to actual management, the employment of bailiffs, safeguarding fish stocks or even augmenting them. Sheffielders wrangled for 40 years about the need to return undersized fish after capture as a breeding stock. The Leeds Angling Association stocked the River Wharfe at Tadcaster with 300 large bream from a Lincolnshire fish farm in 1901, and 500 large perch from Bowness in 1903, as well as placing 1000 perch in the Roundhay lakes, on the city's edge.[68] One relatively rare example of overt paternalism occurred when local professional men helped to stock the Worsley Canal in Lancashire in 1893, offering season tickets at half-a-crown to local working men.[69] It is interesting to contrast the relative ease with which access to fishing was achieved by northern workers' patient negotiation with the prolonged, bitter and occasionally violent struggle to obtain walking rights over nearby moorland which began at the same time and which spilled over into the party political arena in this century.[70] There were no angling 'mass trespasses', if only because the coarse fisherman offered no real threat to the exclusivity of game fishing (once he was kept off it) or to the tradeable values of the waters in which it was pursued. Occasional Sheffield grumbles about restrictions on the Derbyshire waters may have led to occasional poaching, but they did not lead to any overt attempts at invasion. As we shall see, the very nature of match angling effectively bought off the artisan angler.

The amalgamated associations had one other great function, as the wholesale brokers of cheap railway travel, a working man's Thomas Cook. It was widely claimed that the impetus to this was given by a publican, J. Clout of the Berkeley Castle in London's Rahere Street, who founded the Amberley fishing excursions by negotiating a half-crown third-class return fare with the London, Brighton and South Coast Railway Company for his clients in 1870, and using his pub as a ticket office.[71] The capital's West Central Association was prepared, by 1877, to deposit £600 with the Great Western Railway Company to cover a similar concession and negotiated a two-shilling return from Moorgate to Hatfield with the Great Northern Railway.[72] In 1873 the Manchester and District association agreed an eight-shilling return fare from Manchester to Keadby in Lincolnshire.[73] Sheffield, as ever, became the entrepreneur *par excellence*, persuading the Manchester, Sheffield and Lincolnshire Railway Company to offer not just tickets at half-price but to provide special trains as well. On one Monday in 1877, 1740 anglers were ferried out of Sheffield Victoria station, and in 1893 the association ran 120 special trains for 40,000 travellers in the season.[74] It was a sign of the members' affluence, or of their single-minded expenditure on the sport, that they were willing to pay the fares regularly; even a concessionary fare to Woodhall Spa cost 6s 3d, a sixth of a cutler's average weekly wage in the 1900s.[75] The Sheffield Association ran the concession at a profit, to underwrite rental and stocking costs, as well as match prizes. By selling tickets direct, after bulk discounting from the railway, it made £1257 profit on £6000 of trading in 1913.[76] In London the railway companies offered their own version of this when, exasperated by rival bidding from various local associations, they demanded and got one central body with which to negotiate in 1884.[77] In the Midlands control passed into the hands of the Provincial Angling Association, which cornered local concessions from the Great Western and London and North Western Railway Companies.[78] The power to refuse access to these concessions proved to be the strongest sanction available for controlling angling etiquette, as the case of the Clerkenwell Amateurs demonstrated. Powerful they may have become, oligarchic they certainly were, but the associations grew out of club roots and their paternalism was not that of a superior social class, however many honorary vice-presidents some of them chose to nominate.

One strand common to most clubs and associations lay deep in their cultural origins, the inherited mantle of the friendly societies. Most of the pub-based clubs appear to have operated also as benefit societies; so much so that one Hulme angler revealed considerable frustration to the *Northern Angler* in 1893 when, attempting to found a Sunday-school based fishing club, 'pure and simple, i.e. not a benefit society', he had great difficulty in finding a model set of rules without benefit provisions.[79] The few club papers which have passed into archive repositories refer almost invariably to benefits. The Salford Friendly Anglers (Sick) Benefit Society made contributions compulsory for all new fishing members, making every one a sick visitor by rote; illness caused by fighting, wrestling, drunkenness 'or other manifest evil course of life' was not to be recognized.[80] The Warwick Angling Society of Walthamstow stressed that: 'A great thing in favour of the angler is his care for a fallen brother' and charged each member a shilling to help any member's widow when the need arose; the minimum contribution to its benefit fund was one shilling a year.[81] Like many clubs it augmented its relief fund with the proceeds of an annual match, as did the Sheffield Earl of Arundel Philanthropic Society in its fifth Great Annual Match in September 1913.[82] The vicissitudes of working-class life meant persistent and often heavy demands on these funds and the federal associations tried to minimize this by central organization. In London the Anglers' Benevolent Association was formed in 1879, to stamp out what it regarded as a recurrent tax. It charged 1s 1d a year, built links with convalescent homes in Dover and Herne and disbursed quite large sums. In the 1890s, 40 members were sent to the seaside for three weeks entirely free and given a guinea each to spend. But, in attempting to reduce the demands of indiscriminate charity, it firmly eschewed prevalent middle-class models, refusing to publish the names of recipients: 'The committee do not themselves make private enquiries or follow any objectionable Charity Organisation tactics'.[83]

Yet it had recurrent problems; only about 22 per cent of London's anglers joined and only a third of the clubs paid any affiliation, despite repeated appeals.[84] The Sheffield Anglers' Association Sick and Provident Friendly Society fared marginally better from the profits of an annual match.[85] The clubs preferred to look after their own, to go only so far with central bodies. Only in Crewe, Derby and Nottingham did three of these clubs even bother to register with the Registrar General of Friendly Societies; most preferred to exist, as they had always done, in local custom rather than in administrative law.[86] There were times when this sense of local solidarity was employed in meeting wider needs, in the cyclical depressions which hit northern towns. Sheffield was affected in 1878 and the clubs were called on to distribute relief.[87] In 1879 the Manchester Association distributed clothing, food and coals; 'amongst the recipients there were poor anglers who, had it not been for these timely ministrations, must have sunk under the calamity'.[88] In 1893 the Oldham Anglers' Relief Committee distributed 700 two-pound loaves, 700 pints of soup, and tea and sugar from local club rooms.[89] When Sheffield suffered in the same year northern stoicism and the sport's attractions actually contributed to make the impact worse in many homes; 'many a luxury and necessity has had to be foregone to allow of the much-prized excursion being made'.[90] The tensions in this male, work and drink-based sport can usually only be hinted at. Certainly, most associations reported some drops in match attendance in depression years, but it never proved catastrophic.

The immediate and the local continued to serve the majority of needs but the trend towards wider affiliation and a national presence continued. In 1889–90 the northern towns founded the Northern Anglers' Association.[91] Although it had a middle-class element on its committee (including Professor Philip Ransome of Leeds, Arthur Ransome's father, 'a fisherman who wrote history books in his spare time'),[92] it spoke largely for the working-class fisherman, not least in organizing the public meetings which halted the 1896 'rod tax'. It was the Northern Association which convened a meeting in Birmingham in May 1903 at which the National Federation of Anglers (NFA) was formed, immediately attracting eight regional associations with a combined membership of 46,000.[93] The NFA's organization of the first real national match in 1906 represented an attempt to give the sport the structural and ethical weight which the Football Association had established in its Cup Final, with little of the attendant controversy.[94]

IV

Working-class angling was dominated from the mid-Victorian years by competition, the fishing match. So powerful was it in club and association life that it could fairly be claimed that the ordinary pleasure angler had to hide behind the matches to obtain his fishing.[95] The lure of prizes was also credited with the great boom in the sport's popularity among poorer workers in the last decades before 1900. The view that 'there appears something incongruous in fishing against time and tide and weather for a teapot or a meerschaum pipe' was hardly apparent to its devotees.[96]

Prizes were usually on this mundane level, the 'prize copper kettle' beloved of Sheffielders. Little value was attached until quite late to purely honorific rewards such as cups and medals. It is not too fanciful to suggest that the wholesale inclusion in prize lists of useful items such as china, clogs, blankets and sacks of flour had a dual purpose; they made the diversion of household income towards this form of leisure appear less wasteful and possibly tempered domestic resentment of the husband's regular disappearances to the river bank.[97] In this male activity women were rarely mentioned by the angling press, although the big summer matches were widely used as family excursions. With a few exceptions such as a Brothers Well Met annual ladies' match, working-class women did not fish and one of the London associations refused to issue hopeful ones with privilege tickets.[98] The arguments of health and fresh air used to justify upper middle-class women's fly fishing cut little ice in working-class families, although the presence of wives at big matches was claimed to have 'a wholesome restraining influence' on what could easily become a colossal drinking party.[99] Some observers vainly hoped that women's actually joining in would reduce prize-hunting and place 'the art of angling on a higher pedestal', a view which misunderstood the very nature of most of the prizes offered.[100]

The early matches reinforced regional styles. London clubs tended to fish for the greatest number of fish or the greatest weight on a specific day in 'roving' matches, 'to go where they like in public waters', as the Walworth Waltonians did in 1877.[101]

The greater northern difficulties in organizing transport and the progressively greater numbers involved meant that fishing from fixed individual pegs became common early on and was almost universal by the later century, if only to reduce bankside clashes for 'hotspots'.

Matches were useful in bonding clubs together, both internally and in local friendlies, such as when the Lower Broughton Angling Society challenged the Salford Friendly Anglers to a match in 1903; the losers were to pay for the tea at 1s 9d each, no-one to pay for more than two teas.[102] But they also strengthened the whole organizational and economic network of angling. Many of the goods offered in prize lists were donated by local shopkeepers, reinforcing the street links. On a larger scale it emphasized the commercial value of association activities, bringing custom to market towns surrounded by agricultural depression. The tradesmen of Boston, Lincolnshire, marked the importance to them of several Sheffield trips a week in summer by promoting an annual match with £50 worth of prizes and over 500 entrants.[103]

Although the majority of matches were run by the clubs, as when 23 Sheffield clubs ran matches in Lincolnshire on the same day in 1893, a substantial parallel role was played by commercial interests.[104] At the basic level this was usually a publican extending his club links with sweepstakes in which he provided cash and goods prizes, advertising and transport, to be reimbursed by entrance fees ranging from two to ten shillings. One such, fished at Keadby by 300 entrants in July 1878, producing 'an Etna of tobacco smoke', was sponsored by Mr Rotherham of the Grapes Inn, Furnace Hill, Sheffield; he offered £50 worth of prizes in a match fished for six hours, won by a weight of 2lb 7¼ oz.[105] The first prize was often equivalent to several weeks' wages but the attraction of these matches was the sheer range of the prize list: one 1877 Sheffield match offered a first prize of £12 cash and a 50 shilling watch, a second prize of £8 and 63 smaller prizes from 50 shillings downwards.[106] Local Sheffield legend traced prize match origins to a publican, John Wreakes (or Raikes) of the Crown Inn, who sponsored the first one in 1856.[107] In entrepreneurial scale he was soon overtaken by the landlord of the Queen's Head in Scotland Street who started by offering £5 and a live pig in the 1860s, but was providing 100 guineas by 1873 and not less than £100 a match until the later 1880s when his fortunes flagged and his prize input went down to £10. He rarely attracted fewer than 450 entrants a match; his largest ever, in Boston, drew 1300. He was clearly recognized in Sheffield as a 'freelance', outside the growing strength of the Association which tried to exclude him, eventually succeeding with its own 'All England' matches, open to any individual but dominated by locals.[108]

It was a tension repeated elsewhere and only partly resolved when the federal associations took to seeking sponsorship from large firms, particularly breweries, a move which fitted in with centralization in the drink trade. In Manchester one large brewery offered, linked with the local association, a challenge cup, 'open to all bona-fide angling clubs held at any Walker and Homfrey's and Manchester Brewery Co,'s houses'.[109]

In a similar advertising move, the *Hull Times* underwrote matches across the Humber at Brigg in the 1890s.[110]

Organizing matches was far more than a matter of sponsorship; their major bonding role, particularly in the case of the dominant pegged style, was reflected in elaborate rituals for everything but the most informal local friendly competition. Siting the pegs at 8–10 yard intervals was easy enough for a small match but could cover seven miles of bank for a huge sweepstake. The whole day was formalized, from the arrival of excursion trains at the nearest station; in 1894 the 550 competitors and their hangers-on in a Birmingham match were met by a brass band, 'to whose strain the long string of excursionists marched down to the Fleet Inn'.[111]

Pegs were allocated by ballot on the day, in a draw often lasting several hours, after which the trek along the banks began. Fishing often took the shortest part of the day because it was difficult to fish for more than four hours, even in high summer, after travel, a late start and the problems of getting the catches weighed and the anglers back for prize distribution were added up. The beginning and end of fishing were usually signalled by a series of bugle calls by officials stationed along the banks, who watched for malpractice in between.[112] There were endless opportunities for minor self-importance and authority to be exercised.

Disputes arose easily where prizes were concerned. The very smallness of catches in increasingly over-fished rivers and the fine weight margins which separated most prizewinners occasionally encouraged cheating; it was claimed in the 1870s that it was a common

Nottinghamshire habit for 'river sharks' to net coarse fish before a match, hide them in bulky bait carriers and then use them to boost their own catches or sell them to others who wanted to.[113] Most association committees fought a long battle with their members to prevent undersized fish from being used in prize weights and also to ensure the return of fish live to the water after weighing, to preserve stock. They were largely successful by the 1900s, a clear recognition that the catch had passed from the edible to the symbolic in the sport. Match fishing produce ethical debates similar to those common in more athletic sports in the last quarter of the nineteenth century. The 'positive ruffianism of "clubbites" and their persistent slaughter of undersizeable fish' was held by many middle-class pleasure anglers to come from the 'fast-and-loose method pioneered by angling contestants' in the 1870s; if there was a similar reaction from working-class anglers it was rarely heard in such derogatory terms, although club officials worked hard to stamp out 'roughs'.[114]

Their aim was 'a strict and sportsmanlike spirit' and matches proved a major instrument in achieving this by both regulation and example.[115] J.W. Martin, the 'Trent Otter', writing in 1908, saw a wholesale triumph; the 'very scum of the Sheffield dregs' had been forced out, 'no ungentlemanly conduct being tolerated'.[116] It may have been expressed in language more associated with public-school athleticism but it was a working-class voice that spoke it and enforced it. Clubs appear to have become increasingly ready to expel rogue members.

Respectability in match angling never went as far, however, as the obsession with 'amateurism' which pervaded late Victorian track athletics and the football world. Fishing was like bowls in the north of England, with an accepted, if uneasy, closeness between normal pleasure and semi- or wholly professional profit.[117] The place of some prizes in domestic economies has already been noted, although only the most consistently successful could have even recouped their outlay, but sponsorship went hand-in-hand with an extensive and unofficial network of bankside betting. Much of this grew out of street and pub wagers and paralleled the pigeon-racing and shooting and whippet racing of the north. There were a few individuals who became to match fishing what the wagered pedestrians had been to mid-Victorian athletics. In November 1871 a member of the Good Intent Angling Society competed with the Hoxton Brothers (a group of 12–13 year olds from Ye Olde Holywell, Shoreditch – the site of Arthur Morrison's notorious 'Jago') for a £25 side wager.[118] By the 1890s betting was virtually universal and the professionals had moved in, 'attended by all the obvious paraphernalia of the betting ring, with loud-voiced, brass-lunged bookies'.[119]

On the whole it was tolerated, the associations were powerless to do much and they had a fairly clear appreciation of how far they could go on dealing with something which was an intrinsic part of the roots from which their members sprang. Occasionally, there were protests. When the *Hull Times* sponsored a Hull versus Sheffield match at Brigg in 1893, for £10 a side for two teams of ten, a 'Disgusted Hullite' complained that his team was not the town's best, because it was composed of men who fished only for cash. He was supported by the secretary of the Hull Amalgamated who claimed it did not promote 'friendly rivalry among amateurs'.[120] The heroes of this world were not, however, wholly amateurs. Sheffield's Walter Darwin, a persistent winner of the Association's 'All-England' matches (£17 10s and a gold medal in 1895) fished exhibition matches.[121] In 1892 he fished against Joseph Bradshaw of Wigan for £50 in a two-day match split between the Liverpool and Manchester, and Chesterfield and Stockwith, canals, over a total of six hours. It was a blow to Sheffield pride when Bradshaw won, with 102 fish to 20. £250 was estimated to have passed in bets.[122] In February 1894 Darwin took on six customers of the Paradise Vaults, Sheffield for £5 and a supper for the largest number of fish caught in one and a half hours; he took 43 to the others' combined total of 24.[123]

The Sheffield habit of promoting an annual 'All England' match for individuals from the 1870s, with large cash prizes, inevitably fostered this pattern. By 1900 a strong feeling that sporting integrity was being threatened, a definition perhaps of sport for its own sake, saw a slow shift towards team matches, adapted from the custom of many club friendlies. Sheffield offered an annual Haig and Haig Challenge Cup, fished for by 124 teams of four in 1907.[124] Betting went on alongside as usual but at least public probity was observed. When the NFA was formed, 'All England' open matches were effectively replaced by a new national championship. This was sponsored by the *Daily Mirror*, quick to latch on to its readers' prime sporting interests. Affiliated clubs were each allowed to enter a team of ten in the first match in October 1906: 'Eighty Isaak Waltons, under other names' represented 46,000 anglers on the Thames at Pangbourne. It was a contest between regional styles which the Londoners, being on their own waters, expected to win. In the event a Boston team, using Sheffield methods, won and 500 sat down to the post-match banquet where a silver cup and gold medals were presented.[125] Angling had accepted the honorific structure of other Edwardian sports but was still bolstered up by bankside betting, and the semi-professionals could still be found in the succeeding championships held at a different venue each year.

As a sport inseparable from industrial working-class life, match fishing was well established by 1914. Saint Monday died with the First World War but the sport continued to grow, especially as the charabanc augmented the railway trade. Despite the growing popularity of sea-fishing and a dramatic spurt in working-class trout-fishing since the 1960s, it still survives. Bolstered now by television coverage and the sponsorship of high-technology tackle manufacturers it has moved very largely away from its pub and street roots. These have not disappeared entirely, however: Central Television played on the marketability of a pathetic nostalgia when it ran a mini-soap opera series, *Eh, Brian, It's a Whopper!* in 1984.[126] There have been significant structural changes; the once-powerful associations now often provide a cover for the pleasure as often as the match fisherman, dealing as much with individuals as with clubs especially now that the private car has largely replaced collective excursion transport. At the time of writing, in early 1986, the pioneering Sheffielders have begun a strategic retreat from their Lincolnshire waters, unable to compete with other bodies as lease renewals now command inflationary prices, and coarse fishing has become another landowners' revenue-earner.[127] Older fishermen quietly lament this trend and there are still many for whom what was once condemned as the 'wholesale piscatorial butchery' of the fishing match can happily be combined with Izaak Walton's dictum, 'Study to be quiet', with no sense of incongruity, even if they no longer dominate the popular angling world.[128]

NOTES

1. The notable exception is Peter Bartrip, 'Food for the Body and Food for the Mind; the Regulation of Freshwater Fisheries in the 1870s', *Victorian Studies*, 28 (1985): a superb discussion of close season legislation.
2. Based on figures given in the *National Angling Survey 1980*, with suitable extrapolations.
3. *Fishing Gazette (hereafter FG)*, 22 February 1878.
4. D.C. Itzkowitz, *Peculiar Privilege, a Social History of English Foxhunting, 1753–1885* (Hassocks 1977), *passim*; Bartrip, loc. cit., *Fishing Gazette*, 22 February 1878.
5. Charles Kingsley, 'Chalk-stream Studies', *Miscellanies* I (London 1859).
6. Parliamentary Papers, Excise Returns (Parl. Papers 1907, XI) 72–3: (PP 1913–14, XXVI) 766; (PP 1914–16, XXII) 25.
7. Kenneth Robson, 'Izaak's Other River', *Trout Fisherman* (January 1985) 62–3.
8. *FG*, 10 March 1900.
9. Ibid., 24 May 1890 and 11 March 1905.

10. Sheffield City Libraries, *Cuttings*, 12, p. 127, c 1888; *Northern Angler*, 18 February 1893; *FG*, 24 December 1892.
11. T.H. English, *A Memoir of the Yorkshire Esk Fishery Association* (Whitby 1925), p. 22, 28.
12. *Hull News*, 26 April 1884.
13. *FG*, correspondence over October/November 1878.
14. Bartrip, loc. cit., 288ff.
15. West Sussex Records Office, Cobden-Unwin Papers, 966; I am indebted to Sue Millar for this reference.
16. *FG*, 14 March, 18 April 1896; *Sheffield Telegraph* (ST), 27 January 1896; *Sheffield Independent*, 11 May 1896.
17. R.C.Hoffmann, 'Fishing For Sport in Medieval Europe; New Evidence', *Speculum*, 60 (1985).
18. *FG*, 21 July 1878.
19. See my 'Izaak Walton, Father of a Dream', *History Today*, December 1983.
20. *FG*, 25 May 1877.
21. Ibid., 25 February 1893.
22. Ibid., 26 April 1890.
23. J.W. Martin, *Coarse Fish Angling* (London 1908, 1924), p. 21.
24. *FG*, 8 March 1878.
25. Points made by Richard Holt in a paper at the 1985 conference of the British Society for Sports History; see also *FG*, 26 April 1877.
26. *FG*, 21 June 1878.
27. Salford Records Office, U97/07; Sheffield Anglers' Association, *Yearbook*, 1906-07.
28. *FG*, 30 May 1879, 29 August 1878.
29. Ibid., 15 and 22 June 1877, 11 January 1890; W.G. Callcutt, *History of the London Anglers' Association* (London 1923), p. 192ff.
30. Salford Records Office (SRO), U97/5/1.
31. SRO, U97/015; Walthamstow Records Office (WRO); Warwick Angling Society Papers (WAS), *passim*.
32. F.G., 11 January 1890.
33. See my 'Sport and the Victorian Sunday; the Beginnings of Middle-Class Apostasy', *British Journal of Sports History*, 1 (1984).
34. W.G. Callcutt, op. cit., p. 18.
35. *Angler*, 21 July 1984.
36. *FG*, 12 December 1877, 14 January 1878; SRO U97/05/1.
37. *FG*, 29 June 1877, 14 January 1878.
38. Ibid., 14 March 1896.
39. Ibid., 18 January 1878; see D.A. Reid, 'The Decline of Saint Monday, 1766-1876', *Past and Present*, 71 (1976), 76ff.
40. Based on reports in the Sheffield Anglers' Association *Yearbooks* for those dates.
41. Cf. ST, 3 March 1896.
42. S. Pollard, *A History of Labour in Sheffield* (London, 1959), p. 211 and *passim*.
43. M. Wiener, *English Culture and the Decline of the Industrial Spirit, 1850-1980* (Cambridge, 1981).
44. Martin, op. cit., p. 42ff.
45. Ibid., F.G., 4 May 1877.
46. British Association, *Handbook and Guide to Sheffield* (Sheffield 1910), p. 470ff.
47. *FG*, 12, 19, 26 January 1895.
48. Ibid., 11 January 1878.
49. Ibid., 20 August 1892.
50. Ibid., 25 November 1893.
51. Martin, op. cit., p. 18.
52. Arthur Shadwell, *Industrial Efficiency* (London 1905, 1913), p. 110.
53. Ibid. and British Association, op. cit, p. 115; Pollard passes over it, as do other contributors to S. Pollard and C. Holmes, *Essays in the Economic and Social History of South Yorkshire* (Sheffield, 1976).

54. F.G., 26 April 1877, 8 February 1878.
55. Ibid., 13 September 1890; *Sheffield Red Book* (1909), pp. 294–5.
56. *FG*, 18 February 1893, 16 March 1895, 14 March 1896, 4 April 1896, 11 April 1896, 18 April 1896; ST, 5 March 1896; Angler 31 March 1894; *Sheffield Independent*, 15 April; ST, 'Angling Review', November 1962.
57. *FG*, 1 March 1878.
58. Callcutt, op. cit., p. 70ff.
59. Ibid., p. 137.
60. *FG*, 8 March 1890.
61. Callcutt, op. cit., p. 233.
62. *FG*, 12 April 1878, 5 August 1893.
63. Ibid., 26 October 1877.
64. Ibid., 13 June 1879.
65. *Sheffield Mail*, 23 January 1909.
66. Leases held by Sheffield Anglers' Association Ltd.
67. *FG*, 11 February 1893.
68. Ibid., 14 February 1903.
69. *Northern Angler*, 18 February 1893.
70. Cf. my 'Battles for the Countryside' in F. Gloversmith (ed.), *Class, Culture and Social Change; a New View of the 1930s* (Hassocks, 1980).
71. *FG*, 12 April 1878, 20 September 1890; Callcutt, op. cit., p. 20ff.
72. Callcutt, op. cit., p. 29; *FG*, 23 November 1877.
73. SRO, U97/03/1.
74. *FG*, 18 January 1893.
75. Sheffield Anglers' Association, *Yearbook* (1901).
76. Ibid., (1913–14).
77. Callcutt, op. cit., p. 70ff.
78. F.G., 28 January 1893.
79. *Northern Angler*, 25 January 1893.
80. SRO, U97/AB2/3.
81. WRO, WAS, Vol 1, p. 238ff.
82. Sheffield Anglers' Association, *Yearbook* (1913–14).
83. *FG*, 14 November 1903.
84. Ibid., 23 May 1903.
85. Ibid., 10 September 1892.
86. PP 1907 LXVIII, 'Report of the Registrar General of Friendly Societies'.
87. *FG*, 27 November 1878.
88. Ibid., 28 February 1879.
89. *Northern Angler*, 28 January 1893.
90. *FG*, 4 November 1893.
91. Ibid., 8 March 1890.
92. Arthur Ransome, *The Autobiography of Arthur Ransome* (London 1976), p. 24.
93. *The Times*, 19 May 1903; F.G., 23 May 1903.
94. *FG*, 29 September 1906.
95. Bernard Venables, *Fishing* (London 1953), p. 254ff.
96. *FG*, 17 July 1880.
97. SRO, U97/21; F.G., 25 November 1893.
98. *FG*, 22 June 1877, 28 November 1878.
99. *Angler*, 15 September 1894.
100. *Northern Angler*, 3 June 1893.
101. *FG*, 15 August 1879.
102. SRO, U97/C9.
103. *FG*, 30 September 1893.
104. *Northern Angler*, 24 June 1893.
105. *FG*, 12 July 1878.
106. Ibid., 25 May 1877.
107. Ibid., 1 March 1878.

108. Ibid., 6 July 1877; *Sheffield Mail*, 23 January 1909.
109. SRO, U101/A/F1-3.
110. *FG*, 27 August 1892.
111. *Angler*, 15 September 1894.
112. *FG*, 26 April 1877, 13 September 1878.
113. Ibid., 29 June 1877.
114. Ibid., 15 November 1878; Bartrip, loc. cit., 298.
115. *FG*, 22 February 1878.
116. Martin, op. cit., pp. 37–40.
117. Cf. J.M. Pretsell, *The Game of Bowls, Past and Present* (London, 1908), *passim.*; G.T. Burrows, *All About Bowls* (London, 1915), p. 35ff.
118. Callcutt, op. cit., pp. 22–3.
119. *FG*, 14 October 1893; *Northern Angler*, 1 April 1893.
120. *Northern Angler*, 26 August, 2 September 1893.
121. *FG*, 29 July 1893.
122. Ibid., 5, 19 November 1892.
123. *Angler*, 3 February 1894.
124. *Sheffield Yearbook and Record* (1907), pp. 90–1.
125. *FG*, 29 September, 20 October 1906; *Daily Mirror*, 29 September, 15, 16 October 1906.
126. Also published as a book: A. Saville, *Eh. Brian, It's a Whopper!* (London, 1984).
127. *Angling Times*, 22 January 1986.
128. *FG*, 27 September 1879.

7

From Popular Culture to Public Cliché : Image and Identity in Wales, 1890–1914

Gareth Williams

I

Historians are generally distrustful of simplistic generalizations such as that 'the modern age was born in 1900'; and rightly so since it happened in 1905. That year a 25-year-old clerk in a Swiss patent office published three scientific papers which dislodged the two corner-stones of classical physics, the absolute nature of space and time. Albert Einstein's Special Theory of Relativity proposed that space bends: the shortest distance between two points is a curve.

Before the year was out this unlikely hypothesis had received confirmation from an even unlikelier direction. On 16 December, at Cardiff Arms Park, the national rugby team of Wales faced the touring and hitherto invincible New Zealand 'All Blacks'. Some 30 minutes into the game, which would make it around three o'clock, a stratagem which the Welsh players had prepared and practised to a pitch of perfection over the last few days was put to the test. From a scrummage 30 yards outside the All Blacks' goal-line a Welsh attack developed down the right, but the ball was suddenly switched to the left and rapidly passed across field until it reached the diminutive wingman Teddy Morgan, who tore around the converging New Zealand defence on an arcing, curving run that Einstein himself would have approved to score in the corner and win the game.

The following Monday South Wales's leading newspaper carried, amidst acres of match accounts and analyses, an eye-witness description of the occasion in Welsh by one 'Ap Idanfryn', for whom this was plainly a new experience. Whatever his initial reservations, they were in full retreat by the end:

> I had heard about crowds so huge they could not be counted. I saw one today in Cardiff, Wales's principal city, and I have no particular desire to see another like it. Every nation under the sun was represented in this excited throng, black, white, yellow – the

clean-living, rustic Welshmen, the self-confident Englishman, the dark-skinned negro, the yellow-complexioned Jap [*sic*] and Chinese – all assembled to watch thirty strong, rugged men play rugby football . . . Who said the [Welsh religious] Revival has killed off football? There's little to suggest that today, and it has to be admitted that this vast crowd is as intelligent, respectable in character and presentable in appearance as any seen anywhere. How many people are there here? Fifty thousand, they reckon . . . Listen to the cheering. Wales has won! What the other nations of these islands have failed to do has been achieved by these sons of the Ancient Britons. The news is going out to the four corners of the earth. . . . Tonight is going to be as unforgettable as Mafeking night was.[1]

Hailed even at the time as 'the match of the century', the social and cultural symbolic significance of this one rugby game merits as much 'thick description' as any Balinese cock-fight.[2] No play was ever deeper than the Wales versus New Zealand match of 1905.[3]

II

The events of that afternoon would reverberate beyond that year and beyond rugby football itself. The epic nature of the encounter can partly, though by no means wholly, be explained in terms of the aura of mutual invincibility which surrounded it. It was not merely that the Welsh XV – at this time enjoying, like the Wales whose offspring they were, a Golden Age of enterprise, optimism and confidence – had proved superior to the all-conquering New Zealanders who had run up over 800 points against a mere 27 in a whirlwind rampage which had destroyed the cream of the other home countries. It was that both XVs were directly representative of a manner of life as much as a style of play, of a social philosophy as well as of rugby thinking, a classic example of the permeation of a game through the interstices of their respective societies so that it was far more than a game, but a factor defining national existence.

For if both were different societies at the start of this century, they also shared a common desire to define themselves on their own terms to the world at large. Neither was unduly hampered by the stays of tradition and hierarchy which corseted the Mother Country; both were continually fertilizing the technical development of the game by devising new ploys and formations, the 'Celts', the pioneers of the four three-quarter system behind the scrum, the 'Colonials' of a seven-man pack in it. But whereas the Antipodeans would surround their victories with the ideology of the life-enhancing challenge of distant untamed lands, Wales' footballing success was deciphered in other terms, as indisputable, even natural evidence of national progress and, more dubiously, of racial superiority. 'In these latter times Wales has progressed phenomenally' announced its main Liberal newspaper in 1909, cataloguing the prominent places occupied by Welshmen in politics, education, the pulpit, commerce and literature, and careful to add that 'in the higher forms of sport our athletes are at once the envy and the despair of other nations.'[4] The Cardiff *Western Mail*, too, after extracting maximum mileage from the idea of Wales 'coming to the rescue of the Empire' by its victory in 1905, concluded that 'The prestige of Wales has been enhanced tremendously as a nation possessed of those splendid [Celtic] qualities – pluck and determination.'[5] When a new edition of George Borrow's *Wild Wales* appeared in 1906, Theodore Watts-Dunton extolled the historic Welsh characteristics of skill and courage as precisely those qualities possessed by the Welsh team that had just defeated New Zealand.[6]

Such ideas were not merely the stock-in-trade of contemporary racial theorists. They had a special resonance in an Edwardian Wales of political self-confidence, cultural creativity, national self-awareness and material prosperity, all to a greater or lesser degree generated by the produce of a coalfield which was approaching its zenith, diffusing its wealth into all areas of Welsh life and keeping the navies of the world afloat. Earlier in 1905, when the Russian Baltic Fleet was despatched to the bottom of the Sea of Japan, the Kaiser mused that Togo's ships had been fired with Cardiff coal.[7] This Cardiff was the chief funnel-port of the world's greatest single coal-exporting region. It had a third of the entire population of Wales within a 20-mile radius; its own population had shot up from less than 10,000 in 1841 to 164,000 by 1901, and was soon to reach 182,000. This was 'noble and squalid' Cardiff, the Welsh Chicago whose economic and political clout was recognized in October 1905 when it was designated a city, soon to become Lloyd George's 'City of Dreadful Knights'.[8] In December 1905 Lloyd George's own career was launched into the trajectory which would take him to the very pinnacle of power when he was appointed to Campbell-Bannerman's Liberal cabinet. It was Sir Leoline Jenkins, Charles II's Secretary of State (1681–84), who had been the last Welshman to hold such high office; since the later-modern Welsh were as fervently British as their early-modern predecessors, this was an appropriate reassertion of a Welshness that was subordinate to its Britishness. In his celebration of the Welsh victory in cosmopolitan Cardiff in 1905 'Ap Idanfryn' was striking all the dominant key-notes in a Cambro-Imperial fanfare, saluting a Wales confident of its future within the world-wide British polity.

III

It was a Wales ideally respectable, orderly, clean-living, educated, democratic, that was being moulded in its own image by a thrusting middle-class intelligentsia and élite. Their modernization programme involved the reshaping of popular culture, a process which is covered by the umbrella-term 'social control' for the nineteenth century, but which it has become fashionable for early modernists to call 'acculturation'. Historians like R. Muchembled and J. Delumeau intepret the Reformation and Counter-Reformation as a massive missionary crusade aimed at the Christianization of Europe, an attempt to acculturate the masses by their élites.[9] In Peter Burke's definition, this involved 'a systematic attempt by some of the educated . . . to change the attitudes and values of the rest of the population, or as the Victorians used to say, to "improve" them'.[10] The success of the early modern acculturators in turning the inhabitants of Christendom into Christians was, it seems, only partial. *Mutatis mutandis*, if late nineteenth-century rugby football, codified in the English public schools and universities and diffused outwards and downwards, was to play a reforming role in the civilizing of the new Wales of surging demographic and industrial growth, then it cannot be pronounced an unqualified success. It did not curb violence and disorderly behaviour: it incited them. The evidence of Welsh rugby confirms the verdict passed on English and Scottish soccer that 'riots, unruly behaviour, violence, assault and vandalism appear to have been a well-established . . . pattern of crowd behaviour at football matches at least from the 1870s'.[11]

Football teams plausibly provided foci for local identity and fulfilled an important bonding function when the division of labour in industrial society worked in the direction of an autonomous individualism; they block-built identity upwards from the level of works, club and street teams, themselves the formalization of gangs and similar 'abbeys of youth' with long-standing traditions of sociability behind them. But just as the standardization and adoption of a uniform set of rules accelerated the growth of regular fixtures, the increasing involvement of working men brought in its train an increased competitiveness. Proliferating cup and league competitions not only reinforced community identity, they also honed local rivalries to a keen edge. There was 'semi-savagery' and 'wanton biting' in a cup match between Bridgend and Neath in 1880; the winning Llanelli team at Neath in 1883 'were left to the tender mercies of an infuriated mob and had almost to fight their way to the hotel . . . they were hustled and pushed, hooted and pelted with clods, mud and even stones.'[12] Grounds were regularly closed for a specified cooling-off period as the Welsh Rugby Union (founded in 1881 in untamed Neath) took a firmer control of its affairs in the 1890s. Neath's own Gnoll ground was suspended in 1895, Aberavon's in 1896, and Abertillery's in 1897. That year the Rhondda club Ferndale's ground was closed after 'a bloodbath' with adjacent Mountain Ash. Even the Cardiff Arms Park was suspended for five weeks that year because of the manhandling of the referee. In 1899 Aberavon's ground was ordered to be closed for, initially, eight months after the crowd had been incited to violence by a committeeman urging his team to 'go for the [expletive deleted], boys'.[13]

Into the new century, old habits died hard. An incident at Llanelli in November 1900, when the referee had been so intimidated by the crowd that he had required a police escort from the ground and out of the district, led to placards being displayed at Newport urging spectators to accept the referee's decisions, since to do otherwise by shouting and arguing loudly 'is neither fair nor sportsmanlike'.[14] The referee was heeded on a single Saturday in October 1906 when 'discreditable scenes . . . fighting and foul play' resulted in five players being sent off at Mountain Ash and another at Brynmawr, while all 30 players took an early bath at Newport when a harassed referee declared a cessation of hostilities well before time; that day there was 'brutal play' at Cardiff too.[15] Whether this continual mayhem is seen in terms of social scientists' typologies of violence,[16] or as merely the domesticated durability of the love of fighting for fighting's sake, what is clear is that rugby football, one of the new model sports, was not, in Wales at least, played or watched with that devotion by working men to the ideals of sporting behaviour which their middle-class acculturators would have liked. It was not the tranquillizer of unruly proletarian crowds any more than it was the sedative of industrial militancy. In the restless, violent year 1911 enormous crowds thronged to see Wales win its sixth triple crown of the decade, while the domestic Welsh club programme flourished without any diminution of support for its traditional, vigorous rivalries.

IV

It was international success that subdued and eventually reversed the prejudices and, initially, outright hostility of another acculturating agency that did not accommodate within its programme of moral reform anything as conducive to idle living as football, with its attendant evils of drink, gambling and general encouragement to the beast in man – and, it seemed, woman.[17] This was the Chapel. The Church, which had for centuries patronized all kinds of communal events from fund-raising ales to athletic contests, had no such scruples. The Anglican college of St David's, Lampeter, had spearheaded the arrival of rugby in Wales, its peaceful precincts acting as the nursery of tearaway teams that were the very embodiment of the Church Militant; three of the first Welsh representative side were clergymen, including the captain, Bevan; while the captain of the last pre-1914 Welsh XV was another, himself a member of a ferocious pack of forwards known as 'the Terrible Eight'.

The Nonconformist denominations, more puritanical, equated rugby football with all kinds of profanity and intemperance (before the arrival of the modern club house, teams used the local 'Arms' to change in; brewers were traditional sporting patrons). It was not only impressionable young men who were at risk either: a moral minority – probably of one – in Llanelli ringingly declared in 1886 that 'the game of football is unfit for young ladies of both sexes [sic]'.[18] Four years later chapel members at the head of the Swansea valley, having narrowly lost a public debate as to whether rugby should be played in the village, sawed down the goalposts on the eve of the newly-constituted club's first game and solemnly transported them to the police station.[19] At Pontypridd, at the confluence of the major coal-producing Rhondda, Taff and Cynon valleys, where rapid immigration threatened to undermine religion and morality, a minister denounced football as 'the dullest and most senseless game the world has seen . . . If young and middle-aged men wish to frequent pubs, theatres and football fields, then let them in the name of the living God remain outside the Christian pale'.[20] Not even the Sabbath itself was safe: when it became known in 1899 that Swansea intended playing on a Sunday during their forthcoming French tour, local Baptist ministers denounced the proposal as 'an act of public desecration and national retrogression which will injure the morality and fair name of the Principality as the most religious country in the world'.[21]

The Welsh religious revival of 1904–05, which our big-match eye witness 'Ap Idanfryn' had to acknowledge had not killed off rugby, almost did so. A final, defiant gesture against the dying of the light, the anguished cry of a people recently uprooted from its rural verities and thrown into an industrial melting pot – a characteristic product of a frontier society, therefore, uneasily poised between two cultures[22] – the Revival pointed an accusing finger at rugby. As the evangelists crusaded through the coalfield, spectacular conversions were recorded, several, appropriately enough, among the rugby fraternity: followers ostentatiously tore up their season tickets, players burned their jerseys, clubs closed down from anything from four weeks in south east Wales to four years in the more rural, Welsh-speaking anthracite area.[23] The galvanic, if short-lived, effect the Revival had was encapsulated in the conversion of one veteran footballer in mid-Glamorgan who leapt to his feet during a service to proclaim 'I used to play full-back for the Devil, but now I'm forward for God'.[24]

By 1905, however, and certainly afterwards, the future lay with the Devil. It was spectator sport that would now play the role that religion had once enjoyed as a popular mass activity and which the Revival had hoped to regenerate. 'Ap Idanfryn', whose readjustment of his own wrenched sensibility registers like a transplant on a cardiograph, was in truth a Sioni-come-lately by comparison with many who had shared his concerns but who, since the turn of the century, had been persuaded by Wales's striking rugby success to swallow their scruples instead of their blood. 'What is the use of preaching against football?' one deacon 'with a tear in his voice' was heard to remark to another at an international match in 1900. 'We couldn't draw such a crowd as this to a *Gymanfa* [hymn-singing festival].' 'No indeed', replied his companion mournfully, 'but – wasn't that a lovely try!'[25] The Rev. S.B. Williams of Llanelli was, by 1902, condemning 'the attitude of the so-called Christians who objected to going to a football match. He was quite certain that there was less rowdyism among the 12,000 on the occasion of the Swansea match [the speaker was at a club banquet celebrating Llanelli's win] than one would find at a *Gymanfa* or an eisteddfod. Football was a noble game and he did not object etc., etc.'.[26] In 1910 'Play the Game' was the subject of a sermon delivered by a Presbyterian minister in Neyland, Pembrokeshire, in the course of which he rejected 'the rigid system under which he had been brought up [that] had led him to believe that football matches were of such a character that it was wrong for Christian people to support them.' On the contrary, a game of football taught people good temper, and he proposed that local clergymen and deacons formed a team of their own: his preferred line-up would include the Rev. W.J. Chamberlain (Wesleyan) at forward, the Rev. R.C. Evans at three-quarter, the Rev. W. Powell at full-back, and himself –

naturally – in the key half-back position. He also advocated clean play.[27]

It was not only aspirants to religious or moral control who leapt on to the accelerating anti-rugby bandwagon. In 1895 the member for Caernarvon Boroughs wrote to his wife from the industrial valleys of Monmouthshire that his audiences there were less responsive to his radical politics because they were 'sunk into a morbid footballism'.[28] In 1908, when the member for Caernarvon Boroughs was President of the Board of Trade and in Cardiff to receive the Freedom of the City, he saw his first rugby match. He kicked off the Cardiff v. Blackheath game at the Arms Park, and pronounced himself intrigued by it: 'It's a most extraordinary game. I never saw it before and I must say I think it's more exciting than politics'. Five years later his son Gwilym was playing for London Welsh.[29]

Thus were the acculturators themselves acculturated, in accordance with the redefinition of the term 'acculturation' – that first appeared in American anthropological literature around 1880 – by a later generation of anthropologists as embracing 'those phenomena which result when . . . individuals having different cultures come into continuous first-hand contact, with subsequent changes in the original cultural patterns of either or both groups'.[30] The tailoring of a rough and ready popular culture that assimilated a robust physical-contact game like rugby with ease, and fashioned it to suit its own self-making purpose to meet the requirements of more ambitious programmes of reform, resulted in rugby becoming in Wales an annexe of official culture. As a signifier of national identity, rugby had a valuable role to play within the Welsh-British imperial framework.

V

'The men – these heroes of many victories that represented Wales [against New Zealand] embodied the best manhood of the race . . . We all know the racial qualities that made Wales supreme on Saturday . . . It is admitted she is the most poetic of nations. It is amazing that in the greatest of all popular pastimes she should be equally distinguished.'[31] Thus a leading editorial the Monday following the victory which all but stitched rugby football on to the national flag.

Yet this was quite a fortuitous development. Rugby was not intrinsic to the Welsh. How could it be when it was imported into Wales by that familiar diffusionist clutch of headmasters, industrialists, businessmen, solicitors, doctors and clerks? It arrived in Neath, via Llandovery College and Merchiston; Llanelli weathered its early days with the products of Rugby and Wellington at the helm; Newport emerged from a conspiracy of Old Monmouthians now in the professions.[32] Athletic exercise, the paradigm national recreation of Victorian England, was canvassed enthusiastically in Wales. 'To take care and develop systematically our physical condition is not less a duty towards God than ourselves', opined a Welsh sports sage in 1894, 'and the effect is the better development of our mental and moral natures . . . the effect of athletic exercise must benefit the whole of man.'[33] There were close links between athletics and rugby in late nineteenth-century Wales. The Welsh quarter-mile champion died on the rugby field in 1880, while Edward Peake, a member of the first Welsh XV of 1881, had his career curtailed by a hurdling accident at Oxford. Richard Mullock, the Newport impresario who managed that first side, was Wales's representative at the inaugural meeting of the Amateur Athletic Association (AAA) at the Randolph Hotel, Oxford, in April 1880. When, thanks to Mullock, the Welsh Rugby Union was set up the following year, its first president was C.C. Chambers of Swansea, a kinsman of the famous Llanelli-born John Graham Chambers, 'the architect of modern athletics' who during his 40-year life rowed for Cambridge, became champion walker of England, devised the Queensberry Rules, staged the Cup Final and Thames Regatta, founded the Amateur Athletic Club (AAC) in 1866, instituted championships for billiards, boxing and cycling, accompanied (in a boat) Captain Webb as he swam the Channel, and opened a Welsh shop in Chelsea. C.C. Chambers was succeeded as president of the Welsh Rugby Union (WRU) by the Earl of Jersey, a leading South Wales landowner who was for ten years president of the AAA.[34]

While the Welshness of rugby would be retrospectively legitimized by the publication in 1892 of a standard text of George Owen's *Description of Pembrokeshire* (1601), with its bone-crushing account of the Tudor game of cnappan which conveniently endowed the later game with a plausible Welsh–British pedigree, the development of modern rugby owed more to the initiative and resource of immigrants who found the bracing atmosphere of a new South Wales conducive to the expression of various talents, including sporting prowess. These immigrants – a quarter of a million of whom flooded into Glamorganshire alone in the last quarter of the century – made their mark. James Bevan, the captain of Wales's first national rugby side, was actually born in Australia, while the pivotal figure in the introduction of the four three-quarter back-line ('the Welsh system'), was a scion of the Hancock brewing family who left Wiveliscombe in Somerset to assuage Cardiff's thirst in 1883.[35] It then became the conventional wisdom that Welsh rugby success was due to its classlessness, its capacity to transcend social formations and differences by embracing well-educated, white-collar backs (for their skill) and manual worker forwards (for their strength). That forward strength, which was by no means deficient in skill and technical expertise, was supplied by the formidable likes of Packer, Watts, Boucher, Hellings and Brice, who won 60 Welsh caps among them during the years 1891–1907, but whose Welshness was geographical merely: they had all come to work and live in South Wales from Somerset and Devon. Perhaps the first theoretician of Welsh forward play was the brainy Tom Graham who came from Tyneside to win a dozen caps in the 1890s. In this respect even the heroic team of 1905 was more representative of the game in Wales than of some historic ethnicity, and Watts-Dunton, whose facile Darwinism led him to portray them as the reincarnation of the Welsh bowmen of Crecy, might have done better to reflect that 'Boxer' Harding was born in Market Rasen, and H.B. Winfield in Nottingham, while their captain E. Gwyn Nicholls, had been born in Westbury-on-Severn, Gloucestershire. Nor was this unexpected: by 1911 a third of Glamorgan's population was born outside it.

The self-creating Wales which in turn created these rugby heroes had been looking to the outside world since at least the 1880s. That was the decade when a United States consul was appointed to Cardiff and when, in Gwyn A. Williams's words, 'the capital, the technology, the enterprise, the skill and labour of South Wales fertilized large and distant tracts of the world from Montana and Pennsylvania to Chile, Argentina and Russia'.[36] No careers were more global in scope than those of the champion Newport side of the late 1870s, of whom one went to be an army officer in India, another to manage an iron works in Newfoundland, a third to China and a fourth to be a solicitor in Colombo; their captain, W.D. Phillips, went ranching in Texas for 30 years.[37]

This Wales looked out to the world at large and to the Empire in particular. 'Ap Idanfryn' rightly bracketed the jubilation of 16 December 1905 with that of Mafeking night: the Welsh Rugby Union had made several donations in support of the Boer War. Welsh rugby players, especially the professional element among them, are constantly to be found, once their playing careers are over, living – and dying – in the darkest recesses of the Empire. A.J. Gould spent a year in the West Indies in mid-career, while his international brother Bert left for South Africa in 1897 and died in Germiston in 1913. W.B. Norton, who won six caps from Cardiff, died in Old Calabar, West Africa, in 1899; Norman Biggs, also of Cardiff and first capped at 18 in 1888, was fatally wounded by a poisoned dart in Northern Nigeria 20 years later. Strand-Jones of Lampeter and Llanelli, Wales's full back in 1902 and 1903, became a chaplain in Lahore and elsewhere on the Northwest Frontier.[38] The 1905 team, once again, mirror this imperial tendency: Harding emigrated to New Zealand where he became a station-master in Greymouth while J.F. Williams died of blackwater fever in Nigeria in 1911; less exotically in 1910, Percy Bush followed the imperial coal trade from Cardiff to Nantes, where he became vice-consul. Well might *The Welsh Review* of March 1906, after likening the Welsh to the Japanese, rejoice that 'Wales at last is beginning to give of her best to the Empire and the world'.

It was a deeply, or at least sentimentally, royalist Wales. When the Welsh Rugby Union supplanted the South Wales Football Union as the organising body of the game in 1881, it abandoned the former's brash white leek slashed diagonally across a black jersey in favour of the more circumspect Prince of Wales feathers on scarlet. In 1881 Llanelli flirted, fortunately briefly, with a primrose and rose playing strip, a colour combination that might have been suggested by Victoria herself.[39] This was also a profoundly imperial Wales. Even the quasi-separatism of the Young Wales movement, Cymru Fydd, led in the 1890s by Lloyd George and Liberal Chief Whip Tom Ellis, MP for Merioneth, was aimed at winning for Wales a seat in the imperial sun. They both would have endorsed what a Liberal candidate declared to loud applause at Barry on St David's Day, 1900, that 'proud as he was of being a Welshman, he confessed a still greater pride that Wales was part of the British Empire – the largest and best in the world'. This Empire, claimed a Welsh MP and close ally of Lloyd George, was 'wider, more beneficent than that of Rome, more world-embracing than that conceived by Alexander in his wildest dreams'.[40]

Edwardian Wales seemed to have found that much sought-after seat in the sun: 'in these latter times Wales has progressed phenomenally' so that 'Welshmen take their places with the greatest in the sphere of human activity'.[41] Yet educational advance (symbolized by a National University, a National Museum, and, soon, a National Library), political maturity and economic progress, significant though they were, were not the only indicators of a national awakening, as the *South Wales News* recognized in 1909: 'In what may in a sense be termed the Renaissance of Welsh nationalism, football is a factor that cannot be ignored. Wales, in a game that demands more skill than strength – that is, even more a "mental" than a physical pastime – is the equal of, and very often superior to, the other three nations'. It concluded that 'Football is one of the factors that symbolizes individuality in national existence'.[42]

That year Wales won the rugby triple crown for the second

successive year, their fifth of the decade, and they had by now gone
nine international matches without defeat (Australia had been a
further scalp in December 1908). This provided yet further
confirmation, if there was still anybody who doubted it, that rugby
football marked the Welsh out as a distinctive nation. 'The peculiar
qualities of the race may be said to be as distinctly shown in the
exponents of the game as in those who lead in other spheres', and it
was worth re-emphasizing that it had its mental side, and was 'not a
mere exhibition of physical strength or physical skill'.[43]

VI

Imperial Wales' acculturators recognized the potential of rugby as a
focus for national aspiration over and above disruptive political-
industrial tendencies. They not only realized it but almost
marmorealized it, too, when the Herald Bard (T. H. Thomas,
R.C.A.) publicly demanded 'some definite acknowledgement on the
part of the nation generally' of Welsh rugby success, and proposed 'a
memorial column of simple, massive design, treated in the Celtic
style, decorated by discs in bronze'. In suggesting a commemorative
obelisk he was anticipating the future more accurately than he knew,
for the post-war depression would torpedo Welsh prosperity and
confidence on, and off, the field: there would be only one more triple
crown (in 1911) for the next 40 years. The Herald Bard's scheme did
not materialize, but it was not from disagreement with his motives in
proposing it:

> Wales is a very small country. The success which has attended
> her efforts in athletics is therefore a sort of miracle. It has been
> attained by the exercise of those qualities in which critics of the
> Welsh declare us to be deficient – hard work, self-control,
> discipline. The game has been intellectualised by our players.
> Whatever may happen in the future, Wales is signalised.[44]

An image was being manufactured of Welsh rugby far removed from
the reality of the non-ideological excitement and admiration it
evoked, the passions it stirred, its ability to conjure artists out of
artisans, the sheer theatricality which was at the heart of its popular
appeal. Dai Smith has argued that in the face of the destabilizing
combination of continuous in-migration, breakneck economic growth
and the strident politicization of, particularly, coalfield industrial
relations, the concept of Wales as 'an indivisible community, a nation
of natural origins and organic growth' was a difficult one to sustain.[45]
Yet it was strenuously upheld by a two-fold historical presentation of

the inevitable, orderly progress of Wales to its present high plateau of achievement, an improving blend of literary bread and literal circuses. The bread came in the form of a determined output of popular history of the *In the Land of Harp and Feathers* (Alfred Thomas, 1896) variety,[46] the circuses by stage-managed extravaganzas like the Investiture of the Prince of Wales at Caernarvon in 1911, and two years before that at Cardiff Castle, a National Pageant depicting Welsh history from the earliest times. The leading roles were taken by Wales's aristocracy, some of them playing their own forbears, but there was an accompanying cast of thousands – five thousand in fact – so that the Pageant, like Wales itself, could be seen to be embracing 'all classes of the community'. Since this was the function of rugby too, it was altogether fitting that the stars of a scene re-enacting the storming of the Norman-held castle by a local chieftain and his followers were well-known international players, including three of those 'heroes of many victories', Harding, R.T. Gabe and Gwyn Nicholls.[47]

DEWI SANT: Well, well! How my tree has grown! There's glad I am to see it!

10.   Saint David and Dame Wales joyfully contemplate their country's sturdy growth and its expression in the highest forms of human endeavour – the arts, sciences and rugby football (*Western Mail*, 1 March 1910)

The historical and contemporary merging of sporting success and social function made rugby into a badge of national identity in a way that association football, for instance, could not match. Long-established in North Wales through its links with Merseyside, soccer began to make rapid headway in South Wales too from around 1900,[48] but its semi-professional status prevented it from acquiring rugby's cohesive role: it was erroneously but revealingly described in a new journal of 1914 edited by one of Wales's leading figures, Thomas Jones, economist, educationist and later Cabinet Secretary, as 'new and alien . . . the game of the alien of the valleys whose immigration and denationalizing tendency is one of the major problems of our country'.[49] Similarly, attempts to establish professional rugby openly in South Wales in these years ran up against a combination of practical obstacles and social prejudices: grounds proved difficult to acquire and fixtures with professional north of England clubs expensive to maintain, while the proletarian, cloth-cap image of Northern Union (later Rugby League) football worked against securing the support of the influential middle class, and a discriminating popular following could not be weaned from a genuine attachment to the fifteen-a-side game.[50]

To the writer of 1914, however, it was not the intrinsic appeal of the game that explained its popularity among 'the democracy' but that it was an agency of acculturation incorporating the people into an idealized notion of a socially-integrated Welsh nation:

> In the sense that nationality is a community of memories so is Rugby football the national game. The names of the giants are on the lips of the people: there are traditions in Rugby . . . Wales possesses in Rugby football a game . . . which is immeasurably more valuable than the popular code of other countries . . . it has made a democracy not only familiar with an amateur sport of distinguished rank but is in reality a discoverer of democracy which acts as participant and patron.[51]

The fit of image to reality was already under strain before the war, while its aftermath ushered in an era when the optimism, confidence and prosperity which had seemed to gild the Edwardian age were in short supply. By the 1930s it had become, in Wales, commonplace to invoke rugby as the integument that could join together diverse interests across a sorely divided social spectrum.[52] With so many other indicators of a distinctive Welsh identity apparently in full retreat as Wales struggles to survive into the twenty-first century,[53] rugby football retains the image – by now a cliché and often a caricature – it acquired in the early years of the twentieth as the very essence of Welshness.

NOTES

1. *South Wales Daily News (SWDN)*, 18 Dec. 1905.
2. Clifford Geertz, 'Thick Description : Toward an Interpretive Theory of Culture' and 'Deep Play : Notes on the Balinese Cockfight' in his *The Interpretation of Cultures* (London, 1975).
3. The Wales versus New Zealand game is analysed in its total context in D. Smith and G. Williams, *Field of Praise* (Cardiff, 1980), pp. 145–75.
4. *SWDN*, 1 March, 15 March 1909.
5. *Western Mail*, 18 Dec. 1905.
6. T. Watts-Dunton's 'Introduction' to George Borrow, *Wild Wales* (Everyman ed., London, 1906), p. xxii.
7. T.K. Derry and T.L. Jarman, *The European World 1870–1945* (London, 1958), p. 112.
8. For 'the noble and squalid city of Cardiff' see Gwyn A. Williams, *When Was Wales?* (Harmondsworth, 1985), pp. 222–3. More fully, M.J. Daunton, *Coal Metropolis : Cardiff 1870–1914* (Leicester, 1977), and Neil Evans, 'The Welsh Victorian City', *Welsh History Review*, June 1985, 350–87. For an example of the considerable financial clout of the city's business community see A.M. Johnson, *Scott of the Antarctic and Cardiff* (Cardiff, 1984).
9. An excellent conspectus of the literature on social control is provided by F.M.L. Thompson, 'Social Control in Victorian England', *Economic History Review* XXXIV (1981), 189–208. On acculturation, see, e.g. R. Muchembled, *Culture Populaire et Culture des élites dans la France moderne (XVe–XVIIIe siècles)* (Paris, 1978), J. Delumeau, *Le Catholicisme entre Luther et Voltaire* (2nd ed. Paris, 1979).
10. P. Burke, *Popular Culture in Early Modern Europe* (London, 1978), p. 20.
11. J. Hutchinson, quoted in W. Vamplew, 'Sports Crowd Disorder in Britain 1870–1914 : Causes and Controls', *Journal of Sport History*, 7 (Spring 1980), 6.
12. *Bridgend Chronicle*, 2 Dec. 1880; *Llanelly [and County] Guardian*, 6 Dec. 1883.
13. Examples taken from the Minute-Books of the Welsh Rugby Union. Cf. T. Mason, *Association Football and English Society 1863–1915* (Brighton, 1980), pp. 158–67.
14. *SWDN*, 26 Nov. 1900.
15. E.g. Mann and Pearce's FORCE typology, a mnemonic – for frustration, outlawry, remonstrance, confrontation and expressive disorders – adopted by Vamplew, op. cit.
17. Women spectators were well represented at rugby matches from an early date, e.g. 'fully a third' of a 5000 crowd at Llanelli in 1884 'were of the gentler sex', *Llanelly Guardian*, 17 April 1884. 'Football matches are very popular and fashionable places of resort with ladies in other towns', *ibid.*, 25 Nov. 1880.
18. *Llanelly Guardian*, 9 Dec. 1886.
19. *Welsh Rugby Magazine*, Dec. 1970 (feature on Ystradgynlais RFC).
20. *SWDN*, 3 May 1894. The contemporary denominational Welsh language press regularly castigated rugby football as a worthless and dehumanizing activity, e.g. *Y Dysgedydd [The Instructor]* Mehefin [June] 1887, 42–3, *Y Traethodydd [The Essayist]* Gorffennaf [July] 1903, 269–74.
21. *SWDN*, 6 Feb. 1899. There were also protests from the Rhondda and Maesteg, ibid., 12 and 17 Jan. 1899.
22. B.R. Wilson, *Religion in Secular Society* (London, 1966), pp. 27–8.
23. Smith and Williams, op. cit., pp. 126–7. International match attendances were also affected, e.g. *SWDN*, 16 Jan. 1905.
24. Tony Lewis, *The Mules : A History of Kenfig Hill RFC* (Pyle, Mid Glam, 1973), p. 3.
25. *SWDN*, 7 Jan. 1901.
26. *Llanelly Mercury*, 6 March 1902.

27. *Western Mail*, 6 Jan. 1910.
28. K.O. Morgan (ed.), *Lloyd George : Family Letters 1885–1936* (Cardiff, 1973), p. 91.
29. *Western Mail*, 27 Jan. 1908; P. Beken and S. Jones, *Dragon in Exile : The Centenary History of London Welsh RFC* (London, 1980), p. 62. Cf. the Mayor of Cardiff, who attended the Wales v. Scotland match in 1902 not because he knew much about it but 'to give countenance to the national complexion it took', *SWDN*, 4 Feb. 1902.
30. R. Redfield, R. Linton and M.J. Herskovits, 'Memorandum for the study of acculturation', *American Anthropology* 38 (1936), quoted in K. von Greyerz (ed), *Religion and Society in Early Modern Europe 1500–1800* (London, 1984), p. 66.
31. *SWDN*, 18 Dec. 1905.
32. G. Hughes (ed.), *One Hundred Years of Scarlet* (Llanelli, 1983), pp. 1, 11, 51; Nat. Lib. Wales Douglas A. Jones Collection (Neath); W.J.T. Collins, *Newport Athletic Club 1875–1925* (Newport, 1925), pp. 12–13.
33. Hughes, op. cit., pp. 60–1, and generally B. Haley, *The Healthy Body and Victorian Culture* (London, 1978).
34. B. Jarvis, *The Origins of Chepstow Rugby Football Club* (Chepstow, Gwent, 1978), p. 43 (Peake); *SWDN*, 10 March 1880 (Gordon); P. Lovesey, *The Official Centenary History of the AAA* (London, 1979), pp. 19–23 (Chambers, Mullock, Earl of Jersey).
35. J.A. Venn, *Alumni Cantabrigienses* (Cambridge, 1940), Part 2, Vol. i, p. 253 (Bevan); D.E. Davies, *Cardiff Rugby Club History and Statistics* (Cardiff, 1973), pp. 26–8 (Hancock).
36. Gwyn A. Williams, op. cit., p. 222.
37. 'The Record of Newport Rugby' by 'Dromio', *South Wales Argus*, 25 Oct. 1913.
38. *SWDN*, Jan. 1899 (Norton); ibid., 4 March 1908 (Biggs); *The Times*, 10 April 1958 (Strand-Jones); *The Breconian*, Dec. 1911, 54 (J.F. Williams); D.E. Davies, op. cit., p. 65 (Bush).
39. *Llanelly Guardian*, 8 Sept. 1881, 23 Feb. 1882. Hywel Teifi Edwards gleefully explores in Welsh the ultra-royalism of the eisteddfodic élite of Victorian Wales in his *Gwyl Gwalia* (Llandysul, 1980), and in English in *Planet* 52, 12–24. See also D. Tecwyn Lloyd, *Planet* 32, 36–47.
40. *SWDN*, 3 March 1900 (Barry); 24 Nov. 1908 (W. Llewelyn Williams to the Carmarthen Cymmrodorion Society).
41. *SWDN*, 1 March 1909. The Edwardian high-noon is sympathetically assessed by K.O. Morgan, *Rebirth of a Nation : Wales 1880–1980* (Cardiff and Oxford, 1981), ch. 5.
42. *SWDN*, 8 Feb. 1909.
43. Ibid., 15 March 1909.
44. Ibid., 16 and 17 March 1909.
45. Dai Smith, *Wales! Wales?* (London, 1984), p. 51.
46. For this and other examples of a 'Gwaliakitsch of iron-souled feyness that was out to lobotomise all the Welsh on behalf of some of them', see D. Smith, *A People and a Proletariat* (London, 1980), pp. 8–9.
47. *SWDN*, 27 July 1909. Other internationals on view were Charlie Arthur, Alex Bland, Bobby Brice, Billy Douglas, Reggie Gibbs, Billy Spiller and Johnnie Williams.
48. B. Lile and D. Farmer, 'The Early Development of Association Football in South Wales, 1890–1906'. *Transactions of the Honourable Society of Cymmrodorion*, 1984, 193–215.
49. *Welsh Outlook*, Feb. 1914, 18.
50. See my 'How Amateur Was My Valley : Professional Sport and National Identity in Wales, 1880–1914', *British Journal of Sports History*, 2, 3 (1985).
51. *Welsh Outlook*, op. cit., 18–19.
52. When the Welsh national side beat New Zealand again in 1935, the Conservative *Western Mail* (which had in 1928 taken over the Liberal *South Wales Daily News*) hailed it 'as a victory for Wales in a sense that probably is impossible in any other sphere'. See my 'From Grand Slam to Great Slump : Economy, Society and Rugby Football in Wales during the Depression', *Welsh Historical Review* XI (1983), 339–57.
53. An almost apocalyptic gloom pervades the conclusion of Gwyn A. Williams' *When Was Wales?* (1985), pp. 296–306.

PART TWO

CULTURE, SPORT AND 'GREATER BRITAIN'

8

The Pan-Britannic Festival: A Tangible but Forlorn Expression of Imperial Unity

Katharine Moore

I

Throughout the nineteenth century, particularly during the latter decades, the question of the future composition and usefulness of the British Empire was widely debated. The long reign of Queen Victoria had been an era of world domination by the British, a time when the supremacy of the Royal Navy and advances in transportation allowed a tiny island to expand its influence into distant parts of the globe. In the 50 years after Victoria's accession in 1837, one contemporary survey estimated that the area governed by the Queen, exclusive of Great Britain, increased from 1,100,000 to 8,400,000 square miles, the European population of the colonies grew from 2,000,000 to 10,000,000, and the 'coloured' native population rose from 98,000,000 to 262,000,000.[1] This enormous growth had not occurred in accordance with any particular overall plan; much of it had been reactive and circumstantial. But the hindsight afforded Britons at the end of the nineteenth century did allow for a degree of inevitability to be linked to their tremendous record of expansion, often appearing to be a statement of justification. As James Morris described it, 'the separate lines of the Victorian story seemed to have been leading the British inexorably towards the suzerainty of the world . . . their Empire, hitherto seen as a fairly haphazard accretion of possessions, now appeared to be settling into some gigantic pattern'.[2]

However, there were increasing indications that Britain's period of domination was coming to a close. As Kendle has pointed out, the end of the American Civil War in 1865 and the unification of Germany in 1870–71 signalled the onset of a shift in the global economic balance.[3] Both countries industrialized on a massive scale and soon challenged Britain as the world leader in the textile industry, shipbuilding, and machinery production. An alarming military challenge was beginning to emanate from France, Russia and Germany, and even the strength of the Royal Navy was becoming suspect. In spite of these ominous developments, public interest in the Empire had never been higher than in the 1890s. Johnston has offered several reasons for this phenomenon which produced a visible display of public approval for the Empire and its continued existence and potential expansion.[4] The imperialistic enthusiasm shown by some politicians, particularly Lord Rosebery and Joseph Chamberlain, infected the public to a degree. Popular writers, including Rudyard Kipling and H. Rider Haggard, portrayed the essence of the imperial experience as a life of adventure, duty and morality, and the late nineteenth century was also the heyday of imperial writers such as Seeley and Froude. In addition, the establishment of two bodies, the Royal Colonial Institute (in 1868) and the Imperial Federation League (in 1884), helped to promote the spread of British culture to many areas of the world, in large part by presenting the attitude that British civilization was superior to that of all other peoples, as evidenced by the parliamentary and judicial systems, technology, and standard of living enjoyed in the Empire.

The increasing physical size of the Empire, coupled with the slow and often grudging realization that the rest of the world was fast catching up with Britain in economic and military terms, contributed to the steady stream of proposals, both formal and informal, put forward as a means of creating closer ties within the Empire. Some suggestions were political in nature, others involved hypothetical economic and commercial unions and agreements, but all inherently supported the belief that increased links among its constituent members would strengthen the Empire. One author has estimated that at least 150 schemes were debated during the 1870s and 1880s.[5] Of necessity, this chapter must narrow its focus, and therefore the fate of one particular proposal will be examined. The domestic news in Great Britain during 1891 highlighted several important issues, including the increasingly visible suffrage movement, a national election in England, and continuous debate about Home Rule for Ireland. Amid the reporting of these major items a small, anonymous article appeared in the July issue of the periodical *Greater Britain* claiming that the Empire needed a more visible British bond, and suggesting the establishment of a Pan-Britannic Contest of industry, culture, and athletics. This concept captured the attention of prominent men and leading newspapers and periodicals throughout the Empire, and was a regular topic for discussion during the next two years. In time, John Astley Cooper was acknowledged as the originator of the proposal, and his suggestion was seen as an attempt to establish a more tangible unity within the Empire. But by 1894 both the man and the concept had virtually disappeared from public view. The purpose here is to trace the history of the proposed Pan-Britannic Contest, primarily through its contemporary newspaper and periodical coverage, and to uncover the ideals and motives behind this ambitious but ultimately abandoned proposal giving it its due place in cultural history of the British Empire at the end of the nineteenth century.

II

When Cooper's idea first appeared in print in mid-1891, it was immediately seen as something different in that it combined several important aspects of life – culture, industry and athletics – in a grandiose festival celebrating the British race. The concept implied, but did not explicitly state, that the race was superior; Cooper asked if Britons were ready to undertake 'actions for the benefit of mankind, which may make the name of England to be sung for all time as an example to races yet to come?'.[6]

Like many other contemporary proposals, the Pan-Britannic Festival concept sought a tangible way to strengthen links within the Empire; its uniqueness lay in the fact that it included a major sporting component. Cooper saw athletic competition as a vehicle which, with some organizational help, could result in formal bonds being established among the Empire's members. Some sporting contact did exist in the late nineteenth century, and Cooper argued he was merely making use of current conditions to regularize competition in several sports. Perceived changes in the global political and economic balance, increasingly devolved responsibilities in the Colonies with respect to self-government, and the sentiment of a desirable 'family' bond all contributed to the widespread public interest in some sort of federation within the Empire during the latter decades of the nineteenth century. The potential contribution sport could make in this process was significant; as Sandiford has described its role in Victorian society, 'the overriding quest for respectability in the nineteenth century expressed itself in the exuberant manner in which sport was gradually rationalized as the foundation of Anglo-Saxon greatness'.[7] Cooper's personal bias about the importance of athletic competition was stressed clearly from the first. 'Athletic exercises should have a place,' he wrote, 'for before we are a political, or even a commercial and military people, we are a race of keen sportsmen.'[8]

Although relatively few details of the proposed Pan-Britannic Contest had appeared in print, debate on the idea was initiated through J.A. Froude's letter to *The Times* in October in which he praised Cooper's scheme because it was less artificial than other federation suggestions, primarily due to the fact it was attempting to make use of the existing associations in the Empire,[9] particularly sporting links.

It appears that much of the press agreed with Froude's sentiments. *The St. James's Gazette* stressed that a love of sport and the fierce delights of competition was implanted in the Anglo-Saxon breast, and that one of the Empire's strongest links was to be found in common sporting interests: 'Strong is the bond of nationality, strong are the ties of commerce, but stronger than either is the "union of hearts" which comes from devotion to the same forms of recreation'.[10] The South African *Cape Times Weekly Edition* praised the idea as an ideal vehicle for the display of the various resources of the dominions, as well as an opportunity for contests in cultural pursuits and athletic exercises. The paper also judged the concept to be the most practical suggestion for imperial federation so far presented,[11] and subsequently suggested that the possibilities connected with the Pan-Britannic Festival were limitless, much like the potential of the Empire.[12] Various initial reactions to the idea, mostly of a favourable nature, were printed in the home and colonial press, including *The St. James's Gazette*, *Pall Mall Gazette*, *South Africa*, *The Standard*, and *The Times*.

However, it would be misleading to imply that Cooper's proposal met with universal approval and praise. *The Daily Graphic* poked fun at the imagined benefits of the gathering: 'After all, the Olympic games did not prevent the Peloponnesian War; and it is too much to hope that Canadian fishery disputes can be helped to adjust themselves by high-jumps and hurdle races'.[13] A vicious comment on Cooper's idea was printed in *The Saturday Review of Politics, Literature, Science and Art* of 17 October which claimed that the noble idea of imperial federation had been reduced to the level of a picnic every four years, and that Cooper was masquerading as P.T. Barnum trying to organize a gigantic circus.[14] A less blatant but equally pessimistic attack was published in *The National Observer* on 24 October. Among several negative comments on the proposal, the paper praised the sporting aspect of the scheme and stated that 'athletics have done something, and perhaps may do more, towards the consolidation of the Empire'. Sporting tours involving cricket, rugby, association football, and rifle-shooting teams in South Africa, Canada, England and Australia were singled out by the periodical as indicative of current Empire sporting links, providing a strong base from which Cooper's idea might develop.[15]

III

It is evident that Cooper's scheme had caught the interest of many individuals, but he had presented just enough information to cause considerable confusion and speculation as well. On 30 October 1891, *The Times* published a letter in which Cooper sought to outline his ideas in more precise form. Although the athletic portion of the tripartite plan is of the greatest significance in this chapter, the other two components deserve brief attention as important aspects of the total concept. Cooper envisaged the industrial section as a series of conferences among scientific, commercial and industrial representatives of the Empire, the proceedings from which could be presented to the world as a record of progress and as a guide for future development. His belief in the superiority of British life was also evident in the cultural section of the scheme, where Cooper proposed the establishment of national scholarships in science, arts, literature, and technical education, 'open to all enfranchised subjects of the Queen and their families'.[16] It is clear from this excerpt that Cooper meant to include only adult males in the so-called 'White Dominions' of Australia, New Zealand, Canada, and South Africa as well as those subjects eligible in Great Britain. Today, this could be criticized as a racist and sexist view of the world, but those terms were only evolving in Cooper's time (although the Suffrage movement did much to attract attention to gender inequalities), and his proposal represents an outlook commonly held by many British subjects at the close of the last century. Arrogance is a more modern description of this attitude, but the history of the British Empire throughout the nineteenth century provided ample justification, from an English point of view, for Cooper to have presented his idea in a most sincere, if superior, manner.

The detailed plans for the athletic contests reinforced this attitude. The future relationship of the various portions of the Empire, Cooper wrote, rested chiefly in the hands of the young men of the Empire, and an imperial athletic contest would be an ideal vehicle through which to display how secure that future could be.[17] Cooper's initial proposal, which was later expanded, included competition in running, rowing, and cricket, that great imperial link. The question of prizes appropriate for truly amateur competition was of great concern to Cooper, and in this matter his thinking changed dramatically after the initial publication of his idea. In July 1891, Cooper stated that 'the prizes should be in sums of money of great value'.[18] By October he had altered his view significantly, very likely due to a realization that money prizes and amateur athletics were incompatible and unacceptable, and accordingly wrote to *The Times* suggesting that the victors in each athletic event be given some symbolic trophy which would confer fame and honour on the recipient.[19]

Cooper stressed his wish that formality should be kept to a minimum, and that the proposed festival should be staged in a spontaneous and natural way. Here his theorizing breaks down and reveals an impractical ideal. The sheer scope of Cooper's suggestions, involving widely scattered parts of the Empire in a variety of activities, would require a tremendous amount of organization and co-ordination in order to be successfully carried out. The staging of a cultural festival, an industrial exhibition, an athletics meeting, a pageant, and a holiday festival could not be done in a spontaneous way without inviting failure. But Cooper's concern for a free gathering of the family indicates his sensitivity to the difficulties associated with more formal political and commercial unions being mooted at the close of the nineteenth century, proposals which were generally highly artificial in nature and often over-ambitious in design. The case for Cooper's Pan-Britannic Festival was strengthened through his perceptive comment that cultural, industrial and athletic links were already in existence, and he was merely identifying some funding schemes whereby those ties would be made firmer by coming together periodically for a celebration of imperial achievements. But he did not dismiss the potential benefits which could result from these gatherings:

> The scheme involves no political or commercial antagonism, either international or intra-national, while containing tremendous possibilities of political and commercial importance if effected; and by emphasizing the brotherhood of race and promoting sentiment of union it may prepare the way for both closer political and commercial relationship when the colonies are more fully developed.[20]

In the final two months of the year, discussion about the Pan-Britannic Festival appeared in the news on a regular basis. Cooper presented his proposal to the various newspapers and periodicals and invited them to support the concept. Most papers focused on the athletic portion of the proposal, and generally supported the suggestion.[21] A cautious endorsement of the plan was given by *The Midland Evening News*, directly linking Cooper's proposal to that of imperial federation but conceding that the Pan-Britannic Contest stood more chance of coming to fruition because it focused on a 'closer union of sentiment of good will, and good understanding' within the Empire, and was less complex than any political arrangement.[22] An article in *The Colonies & India and American Visitor* carried this thinking one step further by commenting that friendly rivalry was to be encouraged in the Empire, and that, by promoting athletic competition, the way might be prepared for a closer political and commercial relationship 'which we all desire to see, but which none of us at present can see the way clear to bringing about'.[23]

The November issue of *Greater Britain* contained a lengthy article which quoted excerpts from a variety of newspapers commenting on Cooper's idea; it is interesting to note the consistent reinforcement of support for the athletic portion of the gathering and the virtual absence of mention of the other two sections of the proposal, aspects which were nevertheless called equally sociable, if not more useful and educating.[24] The leader itself is noteworthy because it mentions the sporting newspapers only in passing, and does not name any or quote from them, possibly indicating that the sporting press had less status than the dailies, and that support from the leading newspapers would do more to strengthen the case for Cooper's idea than any endorsement from those publications devoted exclusively to sport.

Throughout 1892 it was the athletic portion of the Pan-Britannic concept which received the most attention in the press. This is explained in part by a short article in *The Times* which noted that instead of imposing burdens and conditions, Cooper's proposal would foster a taste which the Anglo-Saxon race in all corners of the world cultivated with enthusiasm. Sporting competition was seen as an important part of British culture, and its role in linking the widely scattered members of the Empire was one which should be retained and expanded. The value of the Pan-Britannic gathering 'might only be sentimental; but those who scan the field of politics, even colonial politics, see that sentiment is still a power in the world'.[25]

During the First World War, Cooper presented a paper to the Royal Colonial Institute in which he revealed more of the philosophy behind his concept of a Pan-Britannic Festival. Patriotically entitled 'The British Imperial Spirit of Sport and the War', the paper began with a flourish:

> I contend, then, that the spirit of sport has sustained many a healthy and productive oasis in a desert of artificialism and commercialism, throughout the British Empire, for many years past. Indeed, the underlying philosophy of all our National and Imperial games is not only to produce skill, discipline, loyalty, endurance, steadiness in attack, patience under misfortune, and other physical and mental qualities, but to encourage unselfish-

ness, which is synonymous with good temper, a sense of humour and honour, as well as healthy hero worship.[26]

It was as if war, like sport, allowed the British spirit the freedom to express itself in a profound way. Cooper's linking of sport and war as parallel experiences was not a novel idea, but his interpretation of the eventual outcome for the Empire was revealing. The Empire, he wrote, had found its soul through the events of the war, the greatest game of the participants' lives.[27] On a practical level, Cooper saw the mingling of the diverse members of the Empire as an excellent opportunity for his ideals to be practised. 'Much mutual misunderstanding, distrust, and ignorance is being cordially destroyed as the men of the British Isles and of the Ocean Commonwealth fight side by side in the trenches of this greater game.'[28]

A letter printed in *The Times* in 1892 from the Hon. James Service, former premier of the state of Victoria, Australia, reinforced the antipodean belief in the strength of sentiment as a bond in the Empire. Service noted that Cooper's idea had obvious and powerful reasons behind it, and anyone who had the permanent unity of the British people at heart should support it.[29] The Hon. J. Ballance, Premier of New Zealand, wrote to Cooper commenting on his proposal, and *The Times* reprinted this correspondence in June the same year. The theme of the value of unity was carried on by the statesman, and he concluded Cooper's games were calculated to establish that unity of sentiment without which no political compact was likely to endure.[30]

Greater Britain continued to be the main promoter of the Pan-Britannic Festival idea in England during 1892. An article in the March issue was typical of the seemingly limitless potential an organized games meeting could hold for the future of the Empire; through friendly sporting competition the constituent members would come to know each other better, and periodic athletic gatherings could pave the way for something larger and more comprehensive, possibly in a political vein.[31] The April and May issues also contained contributions which supported Cooper's proposal and quoted extensively from previously published newspapers and periodicals. The May edition optimistically reported that the scheme had been thoroughly discussed and approved by representative men, statesmen, athletes and educationists, both in the United States of America and the British Empire, and it was now time for the general public to join in the debate.[32] The article clearly stated that the idea had to be endorsed at a higher level, by leading men around the world, before it could enter the domain of the man in the street, providing further insight into the structured and channelled nature of society in late Victorian England. It was a complex social and athletic milieu into which Cooper had launched his proposal.

The British Australasian published a letter written to Cooper by R.P.P. Rowe, past-president of the Oxford University Boating Club, in which the athletic portion of the concept was praised and the superior English attitude in sporting activities was unmistakable. Rowe suggested that the Empire contests should not be held more often than once every four years, so that they would 'not interfere too much with rowing affairs at home'.[33] The Pall Mall Gazette reprinted James Service's letter to The Times and appended to it, in a manner which suggests a 'stamp of approval' had been given, a list of prominent and influential men who had signified their support for the proposed Pan-Britannic idea. The impressive list included Mr Balfour, Earl Spencer, Sir Henry Brougham Loch, the Governor of the Cape, Lord Lansdowne, the Viceroy of India, Lord Harris, Governor of Bombay, Sir Charles Mitchell, the Governor of Natal, Lord Wenlock, Governor of Madras, Sir Walter Sendall and Lord Jersey.[34]

Another lengthy summary article appeared in the June 1892 edition of Greater Britain which hinted at an organized campaign supporting Cooper's concept by stating: 'We made a feature of the proposed Pan-Britannic and All-English-Speaking-Gathering from its inception, with a view of keeping it regularly before our readers'.[35] The magazine also listed more leading advocates of the idea, including the Duke of Fife, the Earl of Derby, Prime Minister Gladstone, and the Prime Minister of New Zealand. A special mention was made of continuing Australian support for the scheme, a unique aspect which deserves some attention here.

Some of the strongest and most loyal support for the Pan-Britannic Festival came from the antipodes. The Commonwealth of Australia was not formally created until 1901, but during the nineteenth century the individual states made considerable progress in economic and political terms, and New South Wales was a good example of emerging prosperity. In 1890, Sydney's largest sporting newspaper, the Referee, hired Richard Coombes, and he worked for the paper until his retirement in 1933. An English immigrant, Coombes dominated the administration of many sports during his lengthy career, including rowing, cycling, coursing, rifle-shooting, and athletics. Coombes was intimately involved with state and national athletics from 1887 to 1935, and his work as a journalist allowed him to promote his personal views in print. By mid-1892 Coombes was a firm advocate of the value of Cooper's Pan-Britannic Festival, and he confidently predicted that athletics would have a more powerful effect upon the Empire in time to come than had been the case for centuries.[36] In spite of the fact that there was very little evidence of concrete movement towards the realization of Cooper's proposal, the *Referee* continued to provide enthusiastic moral support. Although the rate of communication between Great Britain and Australia was relatively slow, Coombes' paper published any items it was sent concerning the Pan-Britannic idea, and he regularly took the opportunity to editorialize on the subject.

IV

The widespread discussion, and in some cases distortion, of his scheme prompted Cooper to publish a lengthy statement in September 1892, finally officially attaching his name to the promotion of the proposal. He remarked that while the idea should not be too closely connected with Imperial Federation, the athletic portion of the concept was designed primarily to raise the standard and objects of athleticism while harnessing the powerful potential of sentiment among the Empire's constituents.[37] Cooper also declared he was not prepared to offer further details until he heard from his unnamed supporters in the Colonies and America. At this stage, Cooper was still soliciting opinions on his proposal outlining an English-speaking gathering of the Empire and the United States.

In October the *Referee* printed an article on the Pan-Britannic Festival which detailed strong American support and connections. Cooper's proposal had remained ambiguous on the question of whether to include representatives from the 'lost' North American colony, and the lack of a consistent name revealed a degree of uncertainty about the exact composition of the gathering. The *Referee* announced it had once again received news that the Amateur Athletic Union of the United States of America was offering to stage the first celebration of Cooper's scheme in connection with the 1893 Chicago World's Fair, and was delighted to report that the proposer welcomed the suggestion. However, one month later the paper reprinted an exchange of letters between Cooper and J.E. Sullivan, Secretary of the American athletic body, which revealed Cooper's response to the invitation as less than enthusiastic. Sullivan believed the World's Fair to be an ideal opportunity for Cooper to put his proposal into place through 'one grand carnival comprising all branches of athletics', but Cooper replied that one of his primary objectives was to host the gathering in England, 'the cradle of the English-speaking race', and thereby politely but firmly refused the offer.[38] So, in spite of the fact that he included the Americans in his plans, Cooper made it clear he wanted competition solely for the Empire as well. A statement reprinted in the *Referee* in mid-December may in part explain this desire. The paper noted that Canada was objecting to American involvement in the Pan-Britannic Games, and that in response, Cooper had proposed that the Colonies and Britain should first engage in a separate competition, with the 'Empire' winners subsequently taking on the Americans.[39]

Greater Britain devoted an entire 31-page issue to 'The Proposed Pan-Britannic and Anglo-Saxon Olympiad' in October 1892. In the introductory article Cooper reiterated his ideas and provided numerous examples of published support for his scheme. The question of leadership was raised once again, but still left unresolved as Cooper wrote: 'My object during the past few months has been to set people thinking about the scheme, and to urge the responsible organisations of Great Britain – Commercial, Educational, and Athletic – to take the idea up, and work it out with their manifold resources'.[40] This excerpt indicates that Cooper expected those men in charge of the various interested bodies to adopt his idea and work out its final details, while it is clear that colonials such as Richard Coombes assumed Cooper, as its originator, would provide the visible leadership for the project. The problem of uncertain leadership would plague the Pan-Britannic idea until its eventual demise. Progress was difficult without practical support from influential circles, and Cooper was either unwilling or unable to gain the necessary backing.

Comment on Cooper's proposal continued to be newsworthy as the new year dawned. *Hazell's Annual* enthusiastically reported that the Pan-Britannic idea had achieved such widespread support that its complete realization was only a matter of time and co-operation.[41] The January 1893 edition of *Greater Britain* contained an anonymous article which employed the previously successful format of quoting positive reports from various newspapers and periodicals; opinions from sources in Australia, Canada, England, South Africa, British Guiana and Malta were indicative of the wide publicity the Pan-Britannic concept received throughout the Empire.[42] But speculation about the exact details of the proposed scheme continued, and Cooper remained reluctant to present any further clarification about the project to the public.

The request for more concrete detail was reiterated by the editor of the *Referee*, and one can sense the frustration of a far-off colony pleading for further information in Coombes' writing: 'What we want is a clearly drawn outline of the scheme. To all intents and purposes we are in accord with the principle of Mr. Cooper's idea'.[43]

Despite a lack of detailed information made available to them, the athletic association in Australia and New Zealand proceeded to plan for a trip to compete in the home country. By May it was reported in the Australian press that Pan-Britannic Games were to be held in London in 1894, and that a representative Australasian athletics team was being formed.[44] In July 1893 the *Referee* reprinted parts of a letter from Cooper to the secretary of the Victorian Amateur Athletic Association in which Cooper stated that South African and Canadian athletes were pledged to come to London in July 1894, and that unfailing support for the competition was now of the utmost importance. But the same paper indicated Coombes' assessment of Cooper's suggestion was somewhat different: 'It appears to me, with all due deference to the promoter, that it is not just enthusiasm that is required just now so much as a definite detailed scheme about which we have become enthusiastic'. In concluding his comments, Coombes also noted that the details of the athletic programme should come properly from the English Amateur Athletic Association (AAA), but apparently no athletic body in England was willing to undertake the practical initiation of the project.[45] Although this excerpt should be regarded primarily as Coombes' personal opinion, it is interesting that the same feeling about the English AAA not doing all it could to promote the Pan-Britannic idea appeared in print in several countries throughout the Empire. The exact nature of the role of the AAA in connection with Cooper's scheme has been difficult to assess, but it is very clear that most Colonial athletic bodies, as well as many in England itself, looked to the AAA for leadership in promoting and establishing the Pan-Britannic Festival.

The minutes of the General Committee of the AAA during the years under investigation contain a few curt and cryptic references to Cooper and his proposal, and the general impression given is one of lack of approval and lack of support. The first mention of the plan was at a meeting on 2 April 1892: 'Mr. Astley Cooper's Pan Britannic contest and festival was before the meeting. It was decided that the A.A.A. could not sanction any contest under A.A.A. Laws between amateurs and professionals'.[46] Cooper's suggestion to allow open competition in athletics obviously did not meet with the approval of the national governing body. The AAA's negative response did not change with the passage of time, for in the next year the minutes recorded that 'a letter was read from the Victorian Amateur Athletic Association on the subject of the Pan-Britannic Festival. The committee decided that until the scheme of the festival was formulated and placed before them it was impossible to take any action in the matter'.[47] Evidently the governing body believed there had been a breach of protocol on Cooper's part.

With many details still left unresolved, Cooper contributed a lengthy article to *The Nineteenth Century* in an effort to settle several lingering questions. He reported that general support for his idea in America, India, Australia, and South Africa prompted him to believe that time and co-operation were all that were needed for a successful staging of the Pan-Britannic Festival to rebaptize the blood bond of Empire members.[48] Cooper devoted two pages of his writing to a discussion of the name of the proposed gathering, but resolved little except to suggest tentatively that the label 'All-Anglian' might be most accurate. More importantly, further details of the scheme were put forward, something many supporters had long desired and requested.

The overall scale of the festival had grown considerably in theory during the two years it had been discussed in the press, and Cooper revealed his plans of what the entire gathering would entail. It was most impressive, and included a conference dealing with matters of social, scientific, commercial, and industrial importance, an athletics meeting, exhibition games of cricket, lacrosse, and baseball, an aquatic contest at Henley, a variety of social festivities and the establishment of scholarships throughout the Empire.[49] The athletic section of the festival had grown tremendously in size since Cooper's initial suggestion of running, rowing and cricket as the only events to be included on the programme. This may reflect increased awareness on Cooper's part of the importance of other indigenous games in certain areas of the Empire, but athletics was to remain the backbone of the sporting section.

In concluding his long article, Cooper reported that his proposal had entered upon its practical stage; a strong and representative committee had been formed in Great Britain, and the idea had found friends among the public men in the Colonies. The idea, in theory, may have been popular, but there is virtually no evidence of practical steps being taken in Great Britain by Cooper or anyone else. One isolated example of concrete support came from Sir E.T. Smith, former mayor of Adelaide, Australia. Smith wrote to Cooper offering financial backing for Australian athletes to travel to the Pan-Britannic Games in London which, according to widespread rumours, were scheduled for 1894. The letter was reprinted in *The Times*, and while the ultimate outcome of Smith's proposal is not known, he at least offered financial support at a time when travel costs loomed as a major obstacle to the successful initiation of the proposed Festival.[50] In Canada, the Amateur Athletic Association discussed the possibility of sending a team of athletes to the proposed meeting, and endorsed the principle of Empire competition very strongly.[51]

While Cooper did not claim his scheme could eternally preserve the

Empire against discordant interests, he hoped it would strengthen the common sympathies and family bond in times of uncertainty.[52] And there were real problems facing the Empire as the nineteenth century drew to a close. Bernard Porter has described the behaviour of Britain in these years as giving a misleading impression, for 'beneath the display there was fear, behind the self-assertion a feeling of vulnerability which could not entirely be hidden'.[53]

Finally, in September 1893, word was received that practical steps had been taken in England to move towards staging the Pan-Britannic meeting. The *Referee* reported that L.A. Cuff, honorary secretary of the New Zealand Amateur Athletic Association, had heard from Cooper that the English AAA had approved his scheme and was ready to work in conjunction with the proposer's committee.[54] No additional sources have been found to support this statement, in fact, no evidence has yet been discovered which indicates any AAA endorsement of Cooper or his ideas. The reported collaboration may have been wishful thinking as far as the involvement of the English AAA was concerned, but Australian support for the idea was very real. However, Coombes was perceptive enough to note that without the co-operation of the various sporting organizations in England, Cooper's scheme would collapse, regardless of support from the Colonies.[55]

The momentum surrounding the Pan-Britannic Festival ebbed as the months of 1894 passed. In spite of a few positive activities such as the formation of the South African Amateur Athletic Association, attributed directly to the promotion of Cooper's idea,[56] few tangible developments occurred which moved the proposal towards fruition. One long-standing problem resurfaced once again in the press: Cooper and the English AAA. The minutes from the General Committee meeting of the AAA on 3 March 1894 provide a hint of underlying ill feeling between the two:

> Owing to a general misapprehension as to the action and position of the A.A.A. in regard to Mr. Astley Cooper's Pan-Britannic Festival, it was unanimously passed 'That a letter be written to the principal London papers over the signature of the Hon. Secretary of the A.A.A. stating that numerous enquiries have been made on the subject of the proposed Pan-Britannic Festival, but that the A.A.A. have been unable to consider the matter as no scheme has been laid before them by its promoters'.[57]

The *Referee* reprinted a letter from Charles Herbert, honorary secretary to the AAA, in which he stated that the organization had been unable to consider the proposed Pan-Britannic Festival because no formal presentation had been made to them.[58] In response, Cooper wrote to *The Times* and took exception to some of Herbert's points; in particular, he made direct reference to his role and that of the AAA, emphasising that his primary objective had been to promote and popularize an idea, and that it was up to the sport's governing body to work out the details of the scheme.[59]

Internal problems in England were a major cause for the stalling of the proposal's progress, but even the faithful *Referee* reluctantly printed a statement by Coombes in October declaring that the Pan-Britannic Olympiad scheme was dead, and that no further effort would be made in Australia to arouse any interest in it.[60] Agreement could be found in Canadian athletic circles as well, where it was reported in late 1894 that the scheme had died a natural death.[61] The final glimpse of the Pan-Britannic Festival proposal in the minutes of the AAA was in December 1894 when a letter from Cooper was read to the meeting;[62] there was no recorded discussion concerning the correspondence, and the matter disappeared from the order of business of the powerful governing body.

V

During the closing decades of the nineteenth century, both in Britain and the Empire, various schemes were put forward which were designed to establish a more formal union in response to perceived changes on the international scene which might threaten Anglo-Saxon superiority. One of these suggestions, unveiled in England in 1891, proposed the establishment of a periodic festival to celebrate the industrial, cultural, and athletic prowess of the race. The idea was one of many designed to strengthen links within the Empire, but its uniqueness lay in the fact that it included a major sporting component. Indeed, the athletic portion soon overshadowed the other two aspects, and Cooper's Pan-Britannic Festival concept was the first large-scale plan of a multisport gathering for the Empire to appear in print. Its unveiling in mid-1891 had brought favourable reaction from the press, and the proposal was a major topic of discussion throughout 1892 and much of 1893. But by 1894 the idea was fading dramatically. Although he had written widely about his proposal, it appears Cooper actually did very little in a practical way to advance his own cause; he clearly saw himself as the inspirational promoter of the Pan-Britannic scheme, not its practical administrator. It was an unhappy circle: supporters in the Colonies constantly asking for more details from the Home Country in order that their own plans might be further advanced; Cooper reluctant to present detailed plans and looking to the various English athletic bodies for actual leadership; and the most powerful sporting organization in England maintaining that proposals had not been properly presented to it so its leadership had not been formally invited. The exact nature of Cooper's relationship with the AAA in particular has yet to be fully clarified, but it appears from the evidence uncovered to date that a certain coolness existed. Had the AAA endorsed or even actively assisted in the promotion of the Pan-Britannic Festival, then success would have been highly likely.

Cooper's idea had a significant amount of theoretical support, but its practical details were never worked out. Cooper did not see this as part of his mandate; unfortunately for the proposed festival, no other group or individual took up that responsibility. So, like many other proposals for closer, more formal links among Empire countries which made headlines during the latter part of the nineteenth century, the idea of the Pan-Britannic Festival shone brightly for a time, and then disappeared. But not without trace. Ironically, Cooper died in February 1930, just six months before the inaugural British Empire Games opened in Hamilton, Canada.

Cooper's unshakeable confidence in the British Empire's remarkable contribution to the development of the civilized world was characterized by a certain religious fervour also displayed by many of his contemporaries. During a period of grave uncertainty in the First World War, Cooper put his belief into words: 'The real secret of the security of the British Empire is that we have acquired it not in the spirit of Hate, or the spirit of Conquest, or the spirit of Greed, but in the genial, good-natured spirit of sport, under which every man must play fair and under which every man gets his chance'.[63] The moral and inspirational role which sport was seen to have in moulding and influencing British and Imperial life was heartily endorsed by Cooper, but he may have unwittingly exposed a major problem: a degree of English apathy towards, and Colonial enthusiasm for, part of the imperial dream. Response from the colonies to the proposed Pan-Britannic Festival was essentially positive, and while the majority of the press coverage in England endorsed the idea in theory, practical support was conspicuously absent. A major stumbling block, probably the most crucial for the festival, was Cooper's relationship with the English AAA. This was characterized by lack of co-operation, a situation which was compounded by an apparent misunderstanding of roles and procedures. On more than one occasion Cooper declared he was only suggesting an idea, and that he expected the appropriate sport governing bodies to work out the practical details; the AAA minutes merely acknowledged Cooper's plans by noting that formal proposals had not been presented to them, thereby assuming their leadership was not required. It was a most unjust fate for an idea which had received a great deal of attention and praise elsewhere.

The appeal of the Pan-Britannic Festival idea probably stemmed from two sources: the novelty of incorporating sporting competition into a suggestion to unite the Empire more formally, as well as the fact that Cooper was proposing to make use of links which already existed in the late Victorian world. The ease with which the athletic component overshadowed, and virtually obscured, the industrial and cultural aspects gives an indication of the popularity of sport, and its potential as a unifying force within the British Empire. Cooper had penetrated to the heart of late nineteenth-century hopes and dreams for the Empire and exposed them through his plans for a Pan-Britannic Festival. But when it mattered most, practical support was not forthcoming in England, and by 1894 a new world-wide athletic festival was being actively promoted by Baron Pierre de Coubertin: a revival of the Olympic Games. Attention was diverted, and the moment for action was lost. More than 30 years passed before the cause was taken up again seriously by another devotee of Empire, Bobby Robinson of Canada. Cooper's proposal to make a contribution to greater unity within the Empire through sport had had a gestation period of almost four decades.

NOTES

1. Lord Brassey, 'Imperial Federation – An English View', *The Nineteenth Century*, Vol. 30, No. 175 (September 1891), 480.
2. James Morris, *Pax Britannica: The Climax of an Empire* (Middlesex, England, 1968), p. 22.
3. John Kendle, *The British Empire-Commonwealth, 1897–1931*(Melbourne, Australia, 1972), p. 2.
4. W. Ross Johnston, *Great Britain Great Empire: An Evaluation of the British Imperial Experience* (St. Lucia, Australia, 1981), pp. 128–9.
5. John Edward Kendle, 'The Colonial and Imperial Conferences 1887–1911: A Study in Imperial Organisation and Politics' (unpublished Ph.D. thesis, University of London, 1965), p. 13.
6. 'Many Lands – One People. A Criticism and A Suggestion', *Greater Britain*, No. 9 (July 15 1891), 462.
7. Keith A.P. Sandiford, 'The Victorians at Play: Problems in Historiographical Methodology', *Journal of Social History*, Vol. 15, No. 2 (Winter 1981), 279.
8. 'Many Lands – One People', op. cit., 461.
9. 'Imperial Federation'. *The Times* (Oct. 14 1891), 6.
10. 'Notes', *The St. James's Gazette* (London), No. 3539, Vol. XXIII, (Oct. 14 1891), 17.
11. 'A Britannic Olympia', *The Cape Times Weekly Edition*, No. 312 (Sept. 16 1891), 17.
12. 'Pan-Britannic Festival', *The Cape Time Weekly Edition*, No. 317 (Oct. 21 1891), 16.
13. *The Daily Graphic*, No. 557, Vol. VIII (October 15 1891), 7.

14. 'Imperial Jinks', *The Saturday Review of Politics, Literature, Science and Art*, No. 1877, Vol. 72 (Oct. 17 1891), 438–9.
15. 'Athletic Pan-Anglia', *The National Observer: A Record and Review*, Vol. VI (New Series), No. 153 (Oct. 24 1891), 574–5.
16. 'The Proposed Pan-Britannic or Pan-Anglian Contest and Festival', Letter to the Editor by J. Astley Cooper, *The Times*, (Oct. 30 1891), 3.
17. Ibid.
18. 'Many Lands – One People', op. cit., 461.
19. 'The Proposed Pan-Britannic or Pan-Anglian Contest and Festival', op. cit., 3.
20. Ibid.
21. As a sample see: 'Proposed All-English Athletic Festival', *The Sporting Life*, No. 5422 (31 Oct. 1891), 8; 'An All-English Festival', *The Daily Telegraph*, (29 Oct. 1891), 7; 'Colonial Gossip by "Outis"', *The British Australasian*, No. 370, Vol. IX (29 Oct. 1891), 1393; 'Mr Astley Cooper's Scheme', *The Colonies & India and American Visitor*, No. 1002 (31 Oct. 1891), Second Edition, 4; *The Argus*, No. 14,150 (31 Oct. 1891), 8; 'Topics of the Day – A "United English Festival"', *The Western Daily Press*, No. 10,420, Vol. 67 (31 Oct. 1891), 5.
22. *The Midland Evening News*, Vol. 15, No. 2379 (Oct. 29 1891), 2.
23. 'Mr Astley Cooper's Scheme', op. cit., 19.
24. 'The Proposed Periodic Britannic Contest and All-English Speaking Festival – A Résumé', *Greater Britain* (15 Nov. 1891), 597.
25. *The Times* (7 Jan. 1892), 9.
26. John Astley Cooper, 'The British Imperial Spirit of Sport and the War', *United Empire*, Vol. VII, No. 9 (Sept. 1916), 581.
27. Ibid., 583.
28. Ibid.
29. 'The Proposed All-English Speaking Gathering', *The Times* (27 May 1892), 4.
30. 'The Pan-Britannic Festival', *The Times* (15 June 1892), 8.
31. 'Britannic Confederation', *Greater Britain* (15 March 1892), 66.
32. 'The Proposed Pan-Britannic and All-English Speaking Festival and Contest', *Greater Britain* (16 May 1892), 142–4.
33. 'Colonial Gossip by "Outis"', *The British Australasian*, No. 394, Vol. X, (14 April 1892), 407.
34. 'Our Kith and Kin', *Pall Mall Gazette* (27 May 1892), 1.
35. 'The Proposed Pan-Britannic and All-English Speaking Gathering', *Greater Britain* (15 June 1892), 169.
36. 'The Proposed Pan-Britannic Festival and Athletic Contest', *Referee*, No. 303 (17 Aug. 1892), 1.
37. J. Astley Cooper, 'An Anglo-Saxon Olympiad', *The Nineteenth Century*, Vol. XXXII, No. CLXXXVII (Sept. 1892), 380–88.
38. 'The Pan-Britannic and Anglo Saxon Olympiad', *Referee*, No. 318 (30 Nov. 1892), 3.
39. 'The Pan-Britannic Games', *Referee*, No. 320 (14 Dec. 1892), 3.
40. 'The Proposed Pan-Britannic and Anglo-Saxon Olympiad', *Greater Britain* (15 Oct. 1892), 278.
41. 'Pan-Britannic and Anglo-Saxon Olympiad', *Hazell's Annual for 1893: A Cyclopaedic Record of Men and Topics of the Day* (London, 1893), p. 525.
42. 'The Pan-Britannic and English-Speaking Olympiad', *Greater Britain* (15 Jan. 1893), 382–4.
43. *Referee*, No. 328 (8 Feb. 1893), 3.
44. *Referee*, No. 343 (24 May 1893), 1.
45. *Referee*, No. 350 (12 July 1893), 3.
46. Minutes of a Meeting of the General Committee of the Amateur Athletic Association, 2 April 1892, Grand Hotel, Manchester.
47. Minutes of a Meeting of the General Committee of the Amateur Athletic Association, 9 December 1893, Guildhall Tavern, Gresham Street, London.
48. J. Astley Cooper, 'The Pan-Britannic Gathering', *The Nineteenth Century*, Vol. XXXIV, No. 197 (July 1893), 81.

49. Ibid., 86–7.
50. 'The Proposed "Pan-Britannic" Gathering', *The Times* (26 July 1893), 9.
51. Minutes of Adjourned Tenth Annual Meeting of the Amateur Athletic Association of Canada Held in the Queen's Hotel, Toronto, Saturday 28 October 1893, 8.
52. J. Astley Cooper, 'The Pan-Britannic Gathering', op. cit., 92.
53. Bernard Porter, *The Lion's Share: A Short History of British Imperialism 1850–1970* (London and New York, 1975), p. 119.
54. *Referee*, No. 361 (27 Sept. 1893), 3.
55. 'The Pan-Britannic Olympiad', *Referee*, No. 369 (22 Nov. 1893), 1.
56. 'The Pan-Britannic and the South African A.A.A.', *Referee*, No. 393 (9 May 1894), 3.
57. Minutes of a Meeting of the General Committee of the Amateur Athletic Association, 3 March 1894, Midland Hotel, Birmingham.
58. 'The English A.A.A. and the Pan-Britannic Festival', *Referee*, No. 400 (27 June 1894), 3.
59. 'The Pan-Britannic Movement', a Letter to the Editor by J. Astley Cooper, *The Times* (15 May 1894), 3.
60. *Referee*, No. 417 (24 Oct. 1894), 3.
61. Hon. Secretary's Report, *Eleventh Annual Report of the Amateur Athletic Association of Canada* Montreal, 29 Sept. 1894, 3.
62. Minutes of a Meeting of the General Committee of the Amateur Athletic Association, 8 December 1894, Grand Hotel, Manchester.
63. J. Astley Cooper, 'The British Imperial Spirit of Sport and the War', op. cit., 593.

9

A New Britannia in the Antipodes: Sport, Class and Community in Colonial South Australia

John A. Daly

South Australia was settled in 1836. Unlike other Australian colonies which had been penal settlements, South Australia was to be 'a new Britannia in the antipodes'[1] – a genuine facsimile of British society. It was to be an experiment in the 'systematic colonisation' ideas of Edward Gibbon Wakefield – 'a miniature England complete in every part according to its proportion',[2] encouraging emigration of all classes and providing 'such social institutions as the settlers (had) been used to in their native land'.[3]

Wakefield was critical of emigration schemes which sought to remove from Britain only those on relief or those convicted of crime. He saw colonization as more than indiscriminate 'shovelling out of paupers' and convicts. Wakefield's theory argued that to 'colonise beneficially' it was necessary to encourage all classes of society 'to emigrate together, forming new communities analogous to . . . the parent state'.[4] It would be 'an entire British community' with all its social ingredients – a landed and merchant class supervising the labouring efforts of a working class within an environment of British institutions. A dynamic equilibrium of land, labour and capital was maintained by a simple solution – land, instead of being given away in the new colony, would be sold at 'sufficient price' to ensure that labourers did not rise 'above their station' too soon after their arrival. Proceeds from land sales would be used to induce others ('suitable young couples') to emigrate with the promise of a free passage. Purchasers of colonial lands would thus be assured of labour to work their properties and would become the gentry of this 'new Britannia'.

South Australia then was to be a Utopia based on Wakefield's principles. The idea of emigration to such a colony proved to be attractive to many who were dissatisfied with life in Britain. Ambitious professionals, younger sons deprived of an inheritance through primogeniture, men of small property, army and naval officers on half-pay and those of 'precarious gentility' struggling to keep up accustomed ways of life as their income deteriorated, were among those who sought opportunities in the new province. So also did religious dissenters who were denied full citizenship rights and religious equality. The new colony promised to separate church and state and this appealed to many middle-class nonconformist townspeople in Britain. These were the free settlers who paid their way to 'the new British province of South Australia'. Wakefield described them as 'the uneasy class' which is probably more explicit than the term 'middle class' in that it describes the frustration of those with education and pride who would, if they could, aspire to higher status. Many of these were to become the gentry of the new colony.

For those who would provide the labour force – the assisted emigrants (or 'poor persons' as they were termed by the Colonisation Commission) – South Australia was publicised as 'a promising, advantageous settlement for the sober, industrious and religious. . . .'[5] and an emigration pamphlet of 1836 warned that free passage would not be available to 'idlers . . . drunkards; but to steady, sober men not ashamed to live by the sweat of their brows . . . (who) cannot fail to become independent in a few years'.[6] This promised reward of independence for industry appealed to many.

During 1836 some 15 ships left England carrying 1,000 emigrants to 'the new British Province of South Australia'. Over the next three years another 10,000 settlers arrived in the colony. The proportion of 'free emigrants' to the 'superior class emigrants' paying their own passage is shown in Table 9.1:[7]

TABLE 9.1

| Year | 'Labouring Class' | | | 'Superior Class' | | |
	Male	Female	Child	Male	Female	Child
1836	433	201	179	78	41	9
1837	383	349	366	88	38	55
1838	900	837	960	258	98	101
1839	1440	1378	1772	373	161	196
1840	786	783	1441	98	25	15
		12,208			1,634	

The journey out from Britain took five months and the memory of that often unpleasant, tiresome sea voyage probably deterred many from going 'home' later when they found pioneering difficult. The new settlers were initially enchanted by 'the paradise' that was South Australia and persisted in describing it in arcadian terms in their letters 'home' while trying to remake it in the image of Britain. The land was neither paradise nor was it England. The climate was different for a start. In fact when the hot summer winds blew from the north it was more like Gehenna than Utopia! They found the heat uncomfortable and inconvenient. Food spoiled. Heavy English clothing was unsuitable. Soft skins were burned, complexions 'ruined'. First attempts at farming were disastrous. Cuttings and seedlings, carefully carried out from Britain, withered and died. The streets of the capital city, Adelaide, had grand names commemorating the founding fathers but they were simply dusty tracks in summer, bogs in winter.

However, despite the hardship and the very different environment, they were eventually able to mould the infant colony into the shape of the society that they had left, or at least a reasonable facsimile. Adelaide was at first no more than a collection of huts and tents, but by 1850 a visitor described it as 'resembling an English town'.[8] The colonists were delighted to receive such comments. Theirs was a conscious effort to develop 'an entire British community'. English trees replaced Australian natives; they planted gardens of lavender and musk, wore English fashions, attended church at eleven. The editor of the *South Australian Gazette and Colonial Register*, George Stevenson, acknowledged the success of this attempt to recreate 'a new Britannia in the antipodes' when he confirmed in an 1845 editorial that: 'English society, manners, language and habits have been successfully transferred . . .'[9]. Sport was one of these 'habits'.

Francis Dutton, describing South Australia for a British audience ten years after its foundation, noted with satisfaction that: 'All the British sports are kept up with much spirit in the colony; hunting, racing and in a less degree cricket are, in the proper seasons much patronised'.[10]

It was true that colonial South Australians sought to express their 'Englishness' through traditional British sports and pastimes. The Mediterranean-type climate encouraged an outdoor lifestyle while the paucity of alternative cultural institutions to occupy leisure time (especially among the 'lower orders') meant that there were few other diversions to engage the colonists' interests.

In spite of concessions that had to be made initially to pioneer living in a very different environment, a colonial gentry *was* evident by the late 1850s and early 1860s. By English standards 'society' was thin in Adelaide but such éliteness only added to the distinction felt and welded the group into a small, cohesive and powerful group. The Adelaide gentry were a clearly defined group. They knew each other well, met at parties and at Government House levees, were often related to each other through marriage.

Their sons were enrolled in a collegiate school, consciously and deliberately modelled on Rugby School in England. Thomas Hughes' book, *Tom Brown's School Days* (1857) was an instant success in Adelaide as it had been in Britain, and it was natural that St Peter's

College, established in 1847 by the colonial gentry, should seek to develop along the lines of the great public schools and encourage sport as 'character-building' in keeping with the Rugby model. George Farr, headmaster of the colonial school from 1854–78, had stroked a college boat at Cambridge and actively encouraged school sport for its 'moral qualities'. When the 'All England Eleven' visited Adelaide in 1877 the whole school received two half-holidays to witness the game and the 'sportsmanship'. In fact Lillywhite's team of professional cricketers opened their Australian tour in Adelaide giving further credence to the belief that South Australia was regarded as 'Britannia in the Antipodes'.

The 'Saints' boys not only played cricket but were encouraged to row, play tennis, attend gym (under the direction of German instructor Adolph Leschen) and compete in athletics (track and field). Leschen argued often that an ideal physical education for boys involved a combination of British games and German 'systematic' gymnastics. However, 'British games' were always popular (and preferred) at Saints by both boys and their parents. When Prince Alfred College was established in 1867 to provide a Methodist alternative to Saints it followed a similar 'muscular Christian' pattern and regularly engaged Saint Peter's College in intercollegiate sporting contests. Both were schools for the sons of Adelaide gentry and sought unashamedly to provide an 'English' education. Even after Adelaide University had opened in 1876 many of the graduates of these Colleges took their degrees at Cambridge or Oxford University.

The daughters of the gentry were either educated at home or attended small private 'ladies' colleges' before 'finishing' their education in England. The best known of the Adelaide girls' schools were the Unley Park School (1855) operated by the Thornber sisters, where cricket was played by the girls before the end of the nineteenth century, Hardwicke College (1882), St Peter's Collegiate School for Girls (1894) and Tormore House (1898) in North Adelaide. This latter school was a preparatory school for university but included callisthenics 'taught by an English physical education mistress' and games (tennis, cricket, hockey (by 1901) and rowing) in the curriculum. The Advanced School for Girls (1879) where Adolph Leschen gave lessons in callisthenics rivalled Tormore for academic excellence and carried 'the imprint of British influence' in emulating English girls' public schools. The sons and daughters of the Adelaide gentry were actively encouraged to pursue sporting interests at school, and joined their parents in select clubs on graduating.

Hunting and horse-racing were the favoured leisure pursuits of the Adelaide 'gentry' as they were of the upper class in Britain. By 1840 there was an Adelaide Hunt Club and the leading colonists (those of

12. Masters and hounds of Adelaide Hunt Club in the late nineteenth century (from John Daly's collection)

11. St Peter's College, Adelaide, v Poonindie (Aborigine) Mission, 1875

'superior class') were members. By the mid-1840s the Hunt Club was holding regular meets each winter. Riders resplendent in traditional hunting garb followed a pack of imported hounds across the plains surrounding the city of Adelaide. In 1847, just a decade after the first landing in the colony, 130 sportsmen turned out for the opening run under the vice-regal patronage of the Governor. The *Register* declared it was 'such an occasion as made them forget they had ever left England'.[11]

However, despite such assertions, there were problems in maintaining an Adelaide Hunt Club, and in the early years of its existence it was often financially embarrassed. There were few in the colony who were prepared (or prosperous enough) to maintain the pack and its servants, and whenever a master returned 'home' to England the Adelaide Hunt Club was under threat. Several times (1850, 1854, 1856) the pack was advertised for sale. The editor of the local paper, *The Register*, described the disposal of the pack and the demise of the Hunt Club as 'disgraceful' in a colony that sought to be a 'real Britannia in the Antipodes'. The absence of this 'truly English sport', he argued, 'could not be tolerated' and he urged the colonial gentry to 'keep the hunt alive in South Australia'. The squire of Fulham, William Blackler, imported a pack of fourteen and a half couple of hounds from his native Devon and re-organized the Adelaide Hunt as a subscription club. Since that date there has always been a hunt club in Adelaide, and indeed it still hunts the hills surrounding the city, albeit following an aniseed drag!

Describing the sports and pastimes of the British in 1869 the Earl of Wilton wrote: 'Let but a few Englishmen assemble in any quarter of the globe and it may be safely predicted that a horse race would be organized. . .'. [12] Within a year of settlement, the first 'Adelaide Races' were held on the extensive plains west of the new town. A contemporary account of that first race meeting held in January 1838 acknowledged that while the dry, dusty 'paddock at Thebarton was far removed from the animation and excitement of Epsom . . . these colonial sportsmen did their best'.[13] To maintain a relationship with 'home' they even ran an annual St Leger and a Derby although the environment and quality of the horses fell far short of their classical counterparts in England.

The 'old English and manly game of cricket' was encouraged quite early in the colony. Anthony Trollope wrote that cricket was the game by which Englishmen might be recognized in every corner of the earth. 'Where a score or so of our sons are found, there is found cricket . . .'[14] An early settler, John Wrathall Bull, pitied the poor misguided young men who needed to chase 'the red ball not over green fields but over dusty paddocks in the heat of the boiling sun'.[15]

But it was British to do so and in 1838 the *Register* invited 'gentlemen cricket players' to attend a meeting at the London Tavern to form a club[16] and arrange matches for the ensuing season.

Obviously the meeting at the London Tavern was successful, for a cricket club was formed, and later that month a further advertisement proclaimed: 'Two members of the London Tavern Club are open to play any two gentlemen a game of single wicket, for ten or twenty pounds.'

Cricket matches attracted vice-regal patronage and 'gentry support'. The leading players in the colony were also leaders in the young community – Edward Gwynne, a solicitor from Sussex who became Attorney-General in South Australia in 1845, Robert Torrens (later Sir Robert), Treasurer and Registrar-General, son of one of the founders of the colony, the sons of Sir Hurtle Fisher, and William Boothby, son of Justice Boothby, are examples. Their efforts were applauded by their elders who endorsed the establishment of cricket in 'the new Britannia' as appropriate and 'civilizing'. It was proclaimed on civic occasions as having a 'wholesome influence' upon social life in South Australia.

However, while cricket was encouraged for all, there were other sports which were exclusive and definitive of the upper class. Genteel South Australians imitated genteel Britons in their leisure activities. Few of the 'great houses' of the Adelaide gentry did not possess a carefully manicured lawn for croquet, described as 'a most infectious' amusement among the upper classes in the 1860s. Expensive sets of balls and mallets were imported from England. When tennis became the fashion in England in the late 1870s Adelaide society adopted the English game and converted the croquet lawn to a court. They played golf and lacrosse, went yachting in the gulf (a Royal Yacht Club was formed in 1846) and imported polo ponies from India. Inside, their homes boasted rooms for billiards and dancing. In the absence of 'proper game' the new gentry imported the fox, the rabbit and the sparrow to provide quarry for the sportsman (and problems for the farmer). They formed exclusive clubs, imported expensive equipment and dressed for the occasions to display their status. In such a manner they consciously strove to be English provincial gentry in colonial South Australia by engaging in symbolic élitist activities. Time-consuming and expensive leisure pursuits such as hunting, yachting, golf and polo all gave opportunities for Adelaide (and South Australian) 'gentry' to dress up and display their status – not only to one another but also to the so-called 'lower orders'.

Sport was not solely the province of the upper class in the new colony. Richard Twopeny who settled in Adelaide in the 1870s and wrote of *Town Life in Australia* observed that 'no class is too poor to play' and added that '. . . the more ample reward attaching to labour out here leaves the colonist more leisure. And this leisure he devotes to working at play'.[17] The inn or tavern provided the initial venue for sport for 'the common man' as it did in Britain. The warm climate encouraged drinking and the pubs were community centres offering recreation and fellowship. Innkeepers acted as entrepreneurs for sporting events, played host to various embryo sporting clubs and gave cover to early bookmakers, for gambling was rife in South Australia as it was in all the colonies. The pubs, in keeping with their British traditions, offered impromptu sporting entertainment for a drinking and gambling clientele. Skittle alleys, quoit grounds, boxing saloons and billiard rooms were frequent additions to Adelaide's early pubs. Some hotels even had shooting galleries or wrestling arenas 'at the back' depending on the sporting interest of the publican. However, cricket and horse races were the common interest of both upper and lower classes. The 'gentry' might own the bloodstock but the common man could enjoy the contest and gamble on the results. When they could not gamble at the races they bet on pedestrians or cyclists. 'Having a bet' was as much sport as the activity itself!

The upper class were openly critical of the gambling of 'the lower orders' but many of the advertised sporting events of the tavern and inn were in imitation of the self-styled 'leaders' of the young community. Shooting, endorsed as another 'popular English pastime', was an activity common to both upper and lower class – only the size of the wager and the venue differed. Pigeon shoots were conducted regularly between sporting gentlemen as the following notice in The *Register* on 8 April 1846 testifies: 'On Monday next . . . a Grand Pigeon Match for forty pounds, between gentlemen of Wiltshire and Sussex will take place on the plains south of Emigration Square'. The 'gentlemen' were not without their imitators for in the *Register* two weeks later Oscar Lines, a prominent sporting publican, informed his clientele that 'a Grand Pigeon Match of twenty members at one pound a member' would take place behind his inn, 'The Halfway House', in the working-class suburb of Hindmarsh.

The 'gambling classes' emulated the gentry in their 'athletic sports'. The Adelaide Amateur Athletic Club, established in 1864 by a group of prominent Adelaide citizens, was followed by a Port Adelaide Club. The gentry applauded the initiative of residents of the working-class area but added that it would be better if events were carried out 'without the adjunct of betting . . .'. Considering the publicized wagers of the upper classes which seemed a concomitant of their sport it is not surprising that their moralistic advice was resented by the Port Adelaide sportsmen.

Middle-class settlers were critical of the sports and pastimes of the working class (but not initially of the Adelaide gentry) particularly because of their association with drinking and gambling and campaigned actively for 'rational recreation'. Organized team games, like cricket and football, flourished under their sponsorship, being justified as both 'manly and moral' and in keeping with the ideal of 'muscular Christianity' prevalent in England at the time. Cricket was 'the game of games' and was always described in eulogistic terms. William Bundey, Attorney-General, defined it as the 'prince of all games'. Delivering a lecture on 'Manly Sports, Exercises and Recreations' to an audience in Adelaide Town Hall in August 1880, Bundey, who, as a successful cricketer and noted yachtsman, practised what he preached, reminded South Australians that the game 'was free to all ranks and upon the field the best man is the one who shows the most skill and endurance'.

Cricket was an integral part of aboriginal 'education'. Bishop Hale introduced cricket to Poonindie mission (a mission station a few miles north of Port Lincoln) in the early 1850s as part of 'the civilizing process'[18] and was pleased to note the pleasure it gave the native boys – and presumably their white onlookers! They were noted for 'not only their neatness in fielding and batting .. . [but] . . . the perfect good humour which prevailed throughout the games: no ill temper shown, or angry appeals. . . .'.[19] Five years later, G. W. Hawkes was able to report in a letter to the editor of the Adelaide *Observer* (18 September 1858) that the Poonindie men had 'played a match with the settlers at Port Lincoln who brought their best players into the field but the natives beat them easily'. They are, he asserted, 'capital cricketers'. In the 1870s the Poonindie Eleven visited Adelaide on a number of occasions and competed ('in appropriate costume') against St Peter's College. Such was the 'value' of the game that it could be played between ancient Australian 'savages' and the sons of colonial gentry! Still, it *is* worth remembering that this fascination with, and success at cricket by, aboriginals was not an isolated case. After all, the first Australian cricket team to tour England (in 1868) was, in fact, made up entirely of aborigines.

'Cricket', wrote Twopeny, 'must take pride of place among Australian sports because all ages and classes are interested in it . . . cricket is the colonial carrière ouverte aux talents.'[20] It was the one game (and perhaps the only institution) where people differently occupied could meet on common ground, where, for a time at least, class distinctions could be ignored. Certainly as village/community clubs proliferated in the latter half of the nineteenth century in South Australia a picture emerges of 'gentry', tradesmen and workers interacting on the cricket ground, their social relationships not obscured by class distinctions. As such, cricket was an integrative sport as opposed to the other exclusive, ritualized display activities of the Adelaide 'gentry'.

Football, which should have been endorsed for the same reasons as cricket, was never ascribed such a high game status. It became the game of masses. Football in South Australia began as a diversion between seasons for gentlemen cricketers. Early players in Adelaide were like counterparts in Melbourne, cricketers of note and men of substance. However, the game ultimely appealed only to the masses because of its vigour, violence and excitement. Concomitantly it lost its appeal to the upper class. Still, despite their opposition, it was noted that the game, like cricket, was a 'muscular Christian' game and both games were consequently endorsed as 'moral and manly' and encouraged for the communal values they stressed and the democracy they fostered. Community teams became foci of local pride and identification. Football lost caste through this popularization[21] although the game continued to be played in 'gentry' schools and by their old scholar teams. Cricket was never 'doubtful'!

Bathing was one activity that presented problems in terms of propriety. The climate, particularly in the summer months, made the pastime attractive and 'rational' yet throughout the nineteenth century bathing was a topic of real concern to those who would preserve the moral standards of this community. A Police Act came into effect in 1845 preventing males from bathing nude in river pools in the vicinity of towns. The upper classes, both men and women, had access to the beaches and could hire bathing machines to ensure privacy and convenience. Coaches ran to the beaches each morning during the summer months and returned to the city in time for businessmen to begin work after a refreshing bathe. However, even at the beaches the sexes were separated – 'ladies to the north (of the jetty) men to the south' – and restricted to early morning or evening hours. Bathers were forbidden by law to 'approach nearer than twenty yards to a person of the opposite sex . . . or nearer than that distance to a bathing reserve of the opposite sex'. Regulations like these were rarely conceived to enhance comfort in the antipodean summer. However, they were an assurance to 'decent folks' that something was being done to protect morals and community standards. The problem was solved by the introduction of the swimming costume and the erection of public baths both at the beach resorts and in Adelaide. Bathing became more popular among all classes as it became more modest and private and a 'professor of the art of natation', Thomas Bastard, or 'Cockney Tom' as he was affectionately known in Adelaide, became lessee of the City Baths and teacher of thousands of South Australian youngsters from 1861.

The South Australian community was not 'entirely British'.

German settlers had arrived in 1838 attracted by the concept of religious freedom. South Australia was widely advertised as a 'paradise of dissent'. A group of approximately 200 immigrants settled on the banks of the Torrens a few miles east of Adelaide, calling their new village Klemzig after their native town in Prussia. Their success enticed others to follow and nearly 7000 arrived between 1838 and 1850. Eventually there were significant German settlements at Hahndorf and Lobethal in the Adelaide Hills and in the Barossa Valley to the north of Adelaide. Initially migration had been for religious reasons but, as time ensued and letters home to Germany advertised the success of early immigrants, the secular incentives of economic improvement and family contact prevailed.

These people formed group settlements and made a real effort to preserve traditional cultural mores and institutions. They spoke German, held annual *Schützenfests* and organized numerous *Vereins* or clubs. The Adelaide *Turnverein* was one of these, and its founder Adolph Leschen played a significant part in establishing Physical Education in South Australian schools by advocating a combination of British games with German gymnastics.[22]

While attempts to preserve German culture (*Deutschturn*) could be construed as isolationist and conducive to friction, this did not occur. There were sporting contests between German and English settlers and indeed cricket became an integrative sport in these communities because that sport was considered an important school game and, even in German community schools, was a popular curricular activity. In Hahndorf, a German town in the Adelaide hills, Douglas Byard was headmaster of an Academy in the latter part of the nineteenth century (1886–1916). An Englishman and an Oxford graduate, he had migrated to South Australia in 1884. He had been a varsity cricketer and introduced the game to the boys at Hahndorf. The 'German school' fielded a team in local competition, and cricket did much to integrate British and German children.

Despite the hope of Wakefield and the liberal theorists who planned South Australia as 'a new Britannia in the antipodes', an 'entire British community', a facsimile of Britain with its rigid class structure, the tendency of colonial life was to annul the prejudices of British society.

The 'leading colonists', the Adelaide gentry, consciously strove to build for themselves a life similar to the one they had aspired to in England. Having achieved wealth and position in the new colony they sought to live the life of English provincial gentry. They built spacious villas around the city and formed literary and scientific associations. The men belonged to the exclusive Adelaide Club and sent their sons to schools in the image of the great public schools of England. They played all the fashionable aristocratic games – polo, lacrosse, archery, golf, tennis. They rode to hounds with the Adelaide Hunt Club and cruised the calm waters of the gulf. They went cycling when it was 'the thing to do'. All were public displays, affirmations of status.

However, the attempts to make actual the image of a colonial aristocracy through such sporting parades were to no avail. The middle and lower classes refused to be impressed. They certainly were not deferential. A utopian ideal of social equality had been part and parcel of the atttraction of emigrating for many of them and consequently they rejected any idea of a colonial aristocracy. Many emigrants to South Australia were radical non-conformists who were sensitive to social differentiation. Their ethos was self-help, self-improvement and respectability. They were hardly likely to be impressed with ostentatious displays of wealth and free time. Anyway by the turn of the century the working classes were quite well off in terms of 'reward for labour' and leisure time. Mark Twain and Anthony Trollope, as visitors to the colony, both assessed it 'a paradise for the working man'. Labour had been in short supply from the start and those who were energetic, showed initiative and were the least bit resourceful did well in the new province. The Saturday half-holiday, achieved in 1865, gave the working man leisure to play games. The eight-hour day awarded in 1873 completed the transition from prolonged labour to reasonable leisure and gave time to practise sport. Trams linking suburbs and trains coupling towns provided communication between communities which sought to pronounce their identity through representative sports teams. The weather was a constant invitation to play and by the turn of the century an observer in South Australia could truly declare that 'the principal amusements of the colonists' (all of them) were 'outdoor sports of one kind or another'.[23]

In a way the roles were now reversed. Veblen's 'leisure class', the Adelaide gentry, were forced by a variety of circumstances (drought, depressions of the 1890s, the growing democratic ethos, technology) to assume a different lifestyle. They could no longer afford conspicuous, time-consuming leisure pursuits as the century closed. They were forced to limit, even decrease, the size of their holdings as difficult economic conditions eroded their wealth. Representation in government and public office had hitherto been a privilege of class. Now it was determined by vote or ability, and if the gentry of Adelaide were not to become 'men of yesterday' but to retain their influence and power they needed to court a new electorate of middle- and lower-class voters who abhorred, it would seem, patrician exhibitions of 'old-country', upper-class sports and games. Their sport became less ostentatious, less public. They could not risk alienating their electorate (South Australia had adult, male suffrage by 1856 – the first Australian colony to do so!) by elaborate displays of British aristocratic sport if they were to remain 'powerful' and retain 'influence'. They now 'served' their electorate by patronage of community sporting associations while pursuing their own sports and pastimes in private. They joined private golf clubs, hunted away from the city in the Adelaide hills across land belonging to sympathetic compeers. They played tennis within the confines of their own homes. Their sons and daughters played sport at school and then either joined their parents' private world of sporting afternoon tea parties or (if they possessed ability) elected to compete with community teams. It was here, at the community level, that sporting prowess was tested, time was taken to hone skills and develop fitness and heroes (and heroines) were determined. Essentially it was now the world of the middle and lower classes for it was they who could afford the time to practise sport, and ability rather than wealth determined success. The youngster of upper-class origins no longer determined who should join his/her company for sport. The 'boot was definitely on the other foot'. The lower orders, if not the new leisure class, were at least the dominant sporting group.

NOTES

1. William Charles Wentworth, *Australasia: A Poem* (London 1823).
2. H.E. Egerton, *Molesworth: Selected Speeches . . . on . . . Colonial Policy* (London, 1913), p. 56.
3. Wakefield quoting the speech of Charles Buller, MP, in his book *The Art of Colonization* (London, 1849).
4. Egerton, op. cit., p. 56.
5. Anonymous, *South Australia in 1842* (London, 1843), p. 32.
6. *The Great Southland*, Emigration Pamphlet, (London, 1836).
7. J.A. Daly, *Elysian Fields: Sport Class and Community in Colonial South Australia* (Adelaide, 1982), p. 18.
8. W.S. Chauncey, *A Guide to South Australia* (London, 1849), p. 61.
9. *Register*, 9 August 1845.
10. F.S. Dutton, *South Australia and its Mines* (London, 1846), p. 145.
11. *Register*, 29 May 184.
12. Earl of Wilton, *On the Sports and Pursuits of the English as bearing upon their National Character* (London, 1869), p. 118.
13. South Australian Jockey Club, *History and Growth of the S.A.J.C.* (Adelaide, 1955), p. 3.
14. Anthony Trollope, *British Sports and Pastimes* (London, 1868), pp. 290, 300.
15. John Wrathall Bull, *Early Experiences of Life in South Australia* (Adelaide, 1884).
16. *Register*, 3 Nov. 1838.
17. Richard Twopeny, *Town Life in Australia* (London, 1833), pp. 204, 202.
18. *Register*, 2 March 1859.
19. John Tregenza, 'Two notable portraits of South Australian aborigines' in *Journal of the Historical Society of South Australia*, No. 12 (1984), 26.
20. Twopeny op. cit., p. 204.
21. Wray Vamplew, *More than fun and games: the historical significance of sport in South Australia* (Adelaide, 1984), p. 12.
22. See John Daly, 'Adolph Leschen: The Father of Gymnastics in South Australia' in *Journal of Historical Society of South Australia*, No. 10 (1982), 92–8.
23. Twopeny, op. cit., p. 204.

10

Latter-Day Cultural Imperialists: The British Influence on the Establishment of Cricket in Philadelphia, 1842–1872[1]

J. Thomas Jable

'Cricket, a sturdy plant indigenous to England; let us prove that it can be successfully transplanted to American soil,' announced Robert Waller in 1843 as he gave a banquet toast following a match between his Philadelphia Union Club and the St George's Club of New York.[2] Waller, an English importing merchant, brought more than merchandise to America. Like most English gentlemen of his era, he believed that sport, particularly cricket, facilitated a young man's transition from youth to adulthood. The cricket field was the great training ground for nineteenth-century English youth. For there they learned to become gentlemen. Cricket requires great personal discipline, co-operation, teamwork and reliance upon others. It takes selective judgement and forethought as well as the development of physical attributes to be successful. It also teaches self-control and sportsmanship – fairness and courtesy to others. In Victorian England, it provided the fibre of the moral fabric which guided Englishmen throughout their adult years. Because cricket worked so well in moulding English gentlemen, the British believed it would be equally beneficial to the less fortunate and unenlightened Americans. Precisely those pretences motivated Waller to promote cricket in America, initially in New York where he helped to found the St George's Cricket Club, and then later in Philadelphia where he organized the Union Club in 1842.[3] Within a generation of the founding of the predominantly English Union Club, cricket had become a permanent fixture in Philadelphia. This essay, then, examines the influence of the British and the impact of other forces that, by 1872, led to the permanent establishment of cricket in Philadelphia.

I

Though Waller's role was crucial to the development of cricket in

Philadelphia, English residents had played the game there as early as 1831. Five years later the game appeared at Haverford College, but soon interest waned and the game all but died there by the decade's end. Although Americans sometimes joined their English brethren at the wicket, they did not play the game in appreciable numbers until the 1840s. The person most responsible for involving native Philadelphians was William Rotch Wister of Germantown. Having mastered townball, the forerunner of baseball, he developed an interest in cricket during his mid-teen years when he observed English textile workers playing it in his neighbourhood. Wister and his younger brothers learned the game from James Thorp, an Englishman who lived near their father's estate. When Rotch Wister, a student at the Germantown Academy, failed in his attempt to introduce the game there, he and his brothers formed a junior cricket club in the early 1840s. They played on the grounds of the Wister estate. But when an errant ball shattered a mansion window, the boys had to look elsewhere for a playing field. Shortly afterwards, Wister's younger brother, John, organized a second junior club in the neighbourhood. It attracted American schoolboys and the sons of English weavers.[4]

While John Wister fuelled the fires of cricket in Germantown, Rotch introduced it to his classmates at the University of Pennsylvania. Shortly after Wister matriculated in 1842, Robert Waller and John K. Mitchell, a professor at the university who had developed an interest in cricket during his studies in England, urged him to organize a junior club for university students. Both gentlemen believed that the younger generation must be involved in order to establish the game in America. Waller proved even more helpful when he permitted the students to use the Union Club's facilities across the river in Camden. The students hired William M. Bradshaw, an English emigrant, as their instructor. In short he instilled in the students 'the proprieties of the game, including obedience to the Captain and submission to the decisions of the Umpire'. Wister maintained that Bradshaw's teachings left an indelible mark on his pupils which contributed to the subsequent 'good order' of Philadelphia cricket. During the early 1850s, however, Waller left Philadelphia, causing the University and Union clubs to disband. Business interests took him back to England and then to New York where he rejoined old friends at the prestigious St George's Cricket Club.[5]

The dissolution of both clubs had a felicitous effect on Philadelphia cricket, for in 1854 Rotch Wister organized the Philadelphia Cricket Club (PCC). He encouraged English residents to join with native

Philadelphians to promote and to preserve cricket. Some Americans may well have adopted the Englishman's moral approbation for playing cricket, but Philadelphians also found other, and perhaps more significant, reasons for playing. 'Our primary idea was exercise and enjoyment in the open air,' revealed Wister, 'and we got both.' In its first season the PCC got little else! Playing three matches against the Washington Club, composed entirely of English players residing in the city's Kensington district, the PCC lost the first two encounters, but salvaged a tie in the third. Each year, as its club membership increased, the PCC expanded its schedule. By 1859 it was playing a ten-match season which included opponents from New Jersey, New York, Maryland, and the District of Columbia. Within four years of its founding, PCC members numbered 106, a 250 per cent increase over the club's 42 charter members. Of these members, though, only a nucleus of perhaps 30 were active cricketers, and of that number, fewer than 20 practised on a regular basis. In fact Wister blamed the first eleven's dismal 0–4 record in 1856 on the members' apathy. 'The result of the matches'. he lamented, was due first to 'a want of prac- tice, and second to a want of practice together as an Eleven'. He then warned that unless the playing members attended practice regularly, 'a different result can scarcely be predicted in future matches'.[6]

In addition to his central role in founding the PCC, Rotch Wister encouraged the youth in his Germantown neighbourhood to organize a cricket club there. The additional club, thought Wister, would provide a local opponent and perhaps even an ideal rivalry for his Philadelphia club. Following his advice, several Germantown lads organized the Germantown Cricket Club (GCC) in 1854, just six months after the PCC had been established. These lads, the younger members of John Wister's junior club, had learned cricket from the Nottingham weavers who had settled in the Germantown area. Though they accepted the help of the English, they were 'thoroughly American in spirit'. The GCC was largely responsible for Americaniz- ing cricket, and it may have been the first club to field an all-native American eleven. The club grew slowly but steadily in the years preceding the Civil War. By then it had achieved notable success with impressive victories over the PCC and other local elevens.[7]

The GCC also triggered, though unintentionally, the formation of the Young America Cricket Club (YACC) in 1855. The older GCC members, unwilling to share playing time with their younger brothers, adopted a policy that prohibited boys under 16 from playing in the matches. Reacting swiftly to the older boys' ban, the enraged youngsters pelted the older boys with apples. This fracas, known as the 'apple barrage', 'caught the attention of Thomas A. Newhall,

Philadelphia's sugar king and father of the famous Newhall cricket clan, and William Wister who fathered Rotch, John, and four other cricket-playing sons. The two gentlemen supported the youngsters' desire to play cricket, and Newhall invited them to play on his property. Soon these young lads formed their own club. From the beginning the Young Americans, unlike their older brothers, vowed to learn the game without the aid of the English residents. They believed cricket would fail in Philadelphia unless they mastered the game themselves. True to their word, they worked diligently on their own and began to make a mark on Philadelphia cricket as America drifted into a civil war.[8]

II

The activities of the Young America, Germantown, and Philadelphia cricket clubs heightened the interest in cricket among Philadelphians. Contributing significantly to that process were the annual English–American matches held at Camden every July during the Independence Day festivities. Initiated in 1856, these matches pitted 18 American cricketers against 11 English residents. The matches became popular local attractions with 6,000 spectators attending in 1858 and 7,000 in 1860. Players from the YACC, GCC, and PCC formed the bulwark of the American contingent. The English players won the first four matches, but when the Americans turned the tables on them in 1860, the *New York Clipper* boastfully called upon the Americans to challenge a British team from England.[9]

What the *Clipper* had conveniently forgotten were the thrashings that British players handed North American teams the previous year. The efforts of Robert Waller, W.P. Pickering of Montreal and Fred Lillywhite of London brought a team of British professionals to North America. The All-England Eleven played matches in Montreal, New York, Philadelphia, Rochester and Hamilton (Ontario) against 22 of the best cricketers in each of those areas. The British team won each match easily; neither Americans nor Canadians could neutralize the play of John Wisden, George Parr, John Lillywhite and Thomas Lockyear. The North Americans' only saving grace was the play of the Philadelphians, who gave the British cricketers their sternest test before succumbing by seven wickets. In fact *Bell's Life*, Britain's sporting newspaper, praised the skill and good fielding of the Philadelphians and acknowledged their unlimited potential, once they developed their bowling. Captain Rotch Wister led the Philadelphia Twenty-two which counted 13 native Americans among its ranks. Even though the Philadelphians imported Harry Wright,

A.H. Gibbes, and six others from clubs in New York and Newark, the locals were pleased with the outcome. They believed they could have done as well without the help of the outsiders. More than 4,000 spectators witnessed the three-day affair held on the renovated cricket grounds at Philadelphia's Camac Woods. This performance not only stimulated great local interest in cricket, but it also moved Philadelphia to the forefront of the game in America. Reporting on the gallant effort of the local cricketers, the *Clipper* proclaimed Philadelphia as 'the fountainhead of cricket' in the United States. And from across the Atlantic, Fred Lillywhite proclaimed that 'cricket in Philadelphia has every prospect of becoming national game'.[9]

Lillywhite's pronouncement of cricket's national stature was indeed optimistic. However, this international match, in conjunction with the English–American Independence Day rivalry and the work of the Philadelphia, Germantown and Young America Clubs, sparked a cricket explosion in Philadelphia. From 1859 to 1861 no fewer than 30 cricket clubs were organized in and around the city. Group after group formed a club; it seemed as if every neighbourhood had one.[11]

Just as Philadelphia cricket entered the national spotlight, the Civil War broke out, curtailing its enthusiasm and disrupting its growth. 'The nursery of American cricket,' lamented the *Clipper*, 'appears to have been materially affected by the troublous state of the times, as they play no away from home matches, and very few at home.' The war, calling young men from playing field to battlefield, decimated the ranks of most cricket clubs. Most of the smaller clubs disbanded, while the larger ones suspended operations due to severe player losses. For instance, seven of the PCC's first eleven joined the Union Army.[12]

In spite of losing its first eleven players to the war effort, the YACC, nevertheless, preserved the game of cricket in Philadelphia. The club remained active because a large number of its members were too young to serve their country in battle. Young boys from several defunct minor clubs joined with the Young Americans to keep the flame of cricket flickering in Philadelphia during its darkest days.[13]

As the tide of Philadelphia cricket receded to its lowest ebb, the game of baseball rushed in rapidly to replace it. But long before baseball invaded Penn's city on the eve of the Civil War, Philadelphians played townball. The Olympic Town Ball Club, the game's chief progenitor, had been in existence since the early 1830s. Most of its members came from the city's merchant and professional classes. The pendulum swung to baseball, however, in 1860 when a team from Brooklyn, the Excelsiors, visited the City of Brotherly Love on its whirlwind tour of the Middle Atlantic region. The

Brooklyn club introduced the New York version of baseball to a composite of Philadelphia townballers representing several local clubs. Though the hometown players easily succumbed to the superior Brooklyn squad, they became enamoured with the 'New York Game'. The Olympic Club immediately turned to baseball, and later that year Philadelphia clubs sent five delegates to the National Association of Baseball Players Convention in New York City. When the convention named one of the Philadelphians as its first vice-president in a gesture of recognition to the Philadelphia delegation, the fate of townball in the Quaker City was sealed.[14] As the baseball mania spread throughout the East, the 'New York Game' began to replace cricket nearly everywhere, but in Philadelphia, it merely surpassed it. Although some local cricketers defected to baseball, England's noble game competed vigorously and successfully with it during the years following the Civil War.

III

When the hope for peace brightened in 1864, cricket reappeared in Philadelphia. That year the PCC played seven matches, of which it won three, and losing twice each to YACC and St George's of New York. It also experienced a resurgence in membership as its roster climbed from 73 in 1862 to 125 in 1865. To improve the skill of its players, the PCC hired a new professional. Down from New York came George Wright who later, with his brother Harry, would attract more attention for his performance on the baseball diamond. Although the PCC's first professional, 61-year-old Englishman Tom Senior, had performed dutifully since the mid-1850s, the club decided to turn to a younger man.[15]

As George Wright helped to bring the PCC to its former stature, two events in 1865 signalled the rebirth of cricket in Philadelphia. The first occurred in June when a composite team of Philadelphia cricketers led by George, Harry and Dan Newhall plus Tom Senior and George Wright defeated a superior team of first-class cricketers from New York City organized by Harry Wright and A.H. Gibbes, two of the best bowlers in America. The second took place three months later when the YACC surprised the all-English St George Club on its home grounds at Hoboken, New Jersey. Not only did the Young Americans defeat one of the best St George's teams in years, but the visitors did it without their three best players. Extolling these recent triumphs of the native Americans, the *Clipper* called for a renewal of the international matches with teams from Great Britain.[16]

While the *Clipper* eagerly awaited the resumption of international competition, the rejuvenation of cricket continued in Philadelphia. The GCC, largely through the efforts of Charles E. Cadwalader, reorganized and moved to new quarters on the estate of Henry Pratt McKean at Nicetown. By 1867 the Germantown club had 212 senior and 23 junior members on its roll. Not to be left out of the post-war cricket revival, West Philadelphians in 1865 formed the Merion Cricket Club. Its membership grew from 15 in its initial year to 61 by 1870. Membership in the PCC climbed to 179 in 1867 as 30 new members were elected in 1866, followed by 44 the next year. The influx of new blood at the PCC carried over to the playing field. Under the direction of a newly-hired English professional, Job Pearson, PCC teams achieved a 14–3 record in 1867. The first eleven won three and lost three, but the second, third, and fourth elevens were undefeated. The following year PCC teams won eight of twelve matches.[17]

The year 1868 marked the return of international competition to the Quaker City. Germantown's Charles E. Cadwalader and his steering committee arranged two matches with the touring British team at the GCC's new grounds at Nicetown. In the first match an All-Philadelphia Twenty-Two, containing nine PCC members, challenged the English Eleven. Led by their captain, George Newhall, the local forces played gallantly in a losing cause, being defeated by just two wickets. The second encounter, however, was not so close. The English visitors overwhelmed an All-United States Twenty-Two by 72 runs. Four years later Cadwalader and his cohorts brought another British troupe to Philadelphia. The English Eleven, headed by the incomparable W.G. Grace, once again accepted the challenge from a local twenty-two. Eleven players in the Philadelphia squad came from the YACC; the GCC provided six and the PCC four. Nearly 7,000 spectators watched the British prevail, but not without a struggle, for the local cricketers played respectably. Henry Chadwick, nineteenth-century sportswriter and cricket buff, classified this match as the finest exhibition of cricket in America.[18]

By 1872 cricket had been firmly established in Philadelphia. Having weathered the period of great oscillation brought about the Civil War, the city's three major cricket clubs – the PCC, the GCC and the YACC – revived local interest in the game. As a result, those clubs experienced significant gains in membership which enabled them to expand their existing facilities or construct new ones. Several new clubs, too, appeared on the scene, bringing the game to the city's expanding territories. In the decade following the Civil War, the YACC dominated Philadelphia cricket largely because the younger members, ineligible for military service, had continued to hone their skills during the Civil War years and because they relied upon themselves to master the game. But in 1872 the club compiled a mediocre 3–2 record. By then its best players had matured and entered the world of business where commitments kept them from playing in most of the matches.[19]

IV

In order to examine Philadelphia's cricket more thoroughly during its formative years, the investigator conducted a socio-economic analysis of the members of Philadelphia cricket clubs in 1860 and 1870. By linking the club members to the *Philadelphia City Directories* and to the 1860 and 1870 *U.S. Population Census Manuscripts*, he obtained information on their age, birthplace, occupation, residence, work-place, real estate and personal property. Of the 201 club members in the 1860 sample, 131 (65.2 per cent) were linked to the census manuscripts, while 174 out of 279 members (62.4 per cent) were linked to the 1870 census. Consequently, there is no information for some of the socio-economic variables for approximately one-third of the club members of each sample. But even for those who were linked to the census, some data are missing for certain variables, particularly real and personal property which sometimes was not reported to the census-takers.

For the 1860 sample, the cases were grouped according to major cricket club (30 or more members) or minor cricket club (fewer than 30 members). This breakdown was selected because the investigators[20] uncovered three major and more than 30 minor cricket clubs in Philadelphia in 1860. By 1870, however, there were four major clubs and just a few minor ones. The investigators could identify only two. Therefore, it seemed most appropriate to examine the 1870 sample according to active and non-active players.

Table 10.1 depicts the ages and birthplaces of the cricket club members. The mean age of the club members for both samples was virtually identical; 28.3 years in 1860 and 28.7 years in 1870. Similarly, the mean ages of the 1860 minor club members (24.9 years) and the 1870 active players (24.4 years) closely approximated to each other. Even more closely related were the average ages of the 1860 major club members and the 1870 non-active players, separated by just one-tenth of a year (29.6 compared with 29.7). These similarities in mean age existed because all the minor club members in the sample were active players and 73 per cent of the major club members were not active as players. The 36 active players in the 1860 major club category had a mean age of 22.0 years.

TABLE 10.1

AGE AND BIRTHPLACE OF MEMBERS OF PHILADELPHIA CRICKET CLUBS IN 1860 AND 1870

| | Age | | Total in sample | Birthplace | | | | Total |
	Mean	Median		Pennsylvania	United Kingdom	Other	Unknown	
1860								
Major clubs	29.6	27.5	102	89 (66.9)*	6 (04.5)	7 (05.3)	31 (23.3)	133 (100.0)
Minor clubs	24.9	23.0	29	17 (23.6)	11 (15.3)	1 (01.4)	43 (59.7)	72 (100.0)
All members	28.3	26.0	131	104 (51.7)	17 (08.5)	8 (03.9)	72 (35.8)	201** (100.0)
1870								
Active players	24.4	23.0	27	25 (53.2)	0 (00.0)	2 (04.2)	20 (42.6)	47 (100.0)
Non-active players	29.7	27.0	147	124 (53.4)	3 (01.3)	20 (08.6)	85 (36.6)	232 (100.0)
All members	28.7	27.0	174	149 (53.4)	3 (01.0)	22 (07.9)	105 (37.6)	279 (100.0)

* Numbers in parentheses indicate percentages.
** Four players were members of major and minor clubs concurrently and were thus counted in both categories.

The overwhelming majority of cricket club members were born in Pennsylvania (See Table 10.1). More than 50 per cent of each sample were native-born. When the unknown cases are eliminated, those percentages climb to more than 80 per cent for both groups. Twelve per cent of the cases in 1860 and nine per cent of them in 1870 were not born in Pennsylvania. This pattern in birthplace substantiated the strong local support cricket had generated in Philadelphia. Somewhat surprising, however, was the paucity of club members born in the United Kingdom. This finding was unexpected because cricket was a British transplant. Though English residents introduced native Pennsylvanians to cricket, they apparently found more comfort in joining cricket clubs organized by fellow Englishmen. Then, too, a number of Britons could have been among the 72 cases that could not be linked to the census manuscripts.

Cricket club members were largely white-collar workers. Table 10.2 reveals that nearly two-thirds of the 1860 sample (132 of 201) and almost seven-eighths of the 1870 sample (241 of 279) fell into the white-collar categories. Within the two general white-collar classifications, the greatest number of cricket club members held high white-collar and professional occupations. In 1860 81 club members (40.3 per cent) worked in high white-collar and professional positions, while ten years later, more than twice that number, 174 or 62.4 per cent, held positions in this category.

Further examination of the high white-collar professional category within the framework of its two divisions – businessmen (merchants, executives and manufacturers) and professionals (lawyers and physicians) – has shown that professionals dominated club memberships in 1860 (50 out of 81; 61.7 per cent), while the merchant/executives surged to the fore in 1870 (107 of 174; 61.5 per cent). This turnabout indicated the increasing popularity of cricket club membership for Philadelphia businessmen.

Very few club members worked in blue-collar occupations. Of the 35 club members holding blue-collar positions in 1860, 33 belonged to minor clubs. That represented 45.8 per cent of the minor clubs' memberships; those minor club members working at white-collar jobs were close behind at 41.7 per cent. In spite of the blue-collar workers who were involved with the minor clubs, white-collar workers dominated the cricket clubs in 1860 and 1870. By the latter year, blue-collar workers had all but disappeared from club membership rosters. Only three blue-collar workers appeared in the 1870 sample. From these data, it is clear that cricket club members, for the most part, were white-collar workers. This finding, though not unexpected, nevertheless, corroborates the qualitative literature on Philadelphia

TABLE 10.2

OCCUPATIONS OF MEMBERS OF PHILADELPHIA CRICKET CLUBS IN COMPARISON WITH NATIVE WHITE AMERICANS FOR 1860 AND 1870

	Blue collar	Low white collar	High white collar and professional	Student	Unknown	Total
1860						
Major clubs	3 (02.3)*	31 (23.3)	75 (56.4)	0 (00.0)	24 (18.0)	133 (100.0)
Minor clubs	33 (45.8)	20 (27.8)	10 (13.9)	1 (01.4)	8 (11.1)	72 (100.0)
All members	35 (17.4)	51 (25.4)	81 (40.3)	1 (00.5)	33 (16.4)	201 (100.0)
Native white Americans**	7196 (58.6)	2800 (22.8)	786 (06.4)	–	1498 (12.2)	12,280 (100.0)
1870						
Active players	0 (00.0)	9 (19.1)	28 (59.6)	6 (12.8)	4 (08.5)	47 (100.0)
Non-active players	3 (01.3)	58 (25.0)	146 (62.9)	13 (05.6)	12 (05.2)	232 (100.0)
All members	3 (01.2)	67 (24.0)	174 (62.4)	19 (06.8)	16 (05.7)	279 (100.0)
Native white Americans	6541 (60.4)	2708 (25.0)	574 (05.3)	–	1007 (09.3)	10,830 (100.0)

* Numbers in parentheses indicate percentages.
** Data on native white Americans are from the Philadelphia Social History Project data base and appear in Scott C. Brown, 'Migrants and Workers in Philadelphia: 1850 to 1880,' Ph.D dissertation, University of Pennsylvania, 1981.

cricket which depicts most of the club members in the white-collar occupational brackets.

In order to determine the relationship of the cricket club members from an occupational standpoint to the rest of Philadelphia society, the investigator compared the club members with the Philadelphia Social History Project's sample of native-white Americans. Table 10.2 shows that the club members were disproportionately represented in the high white-collar and professional category (40.3 per cent against 06.4 per cent in 1860; 62.3 per cent against 05.3 per cent in 1870). The increase in the percentage of club members holding high white-collar and professional positions in 1870 was due chiefly to the preponderance of businessmen who joined cricket clubs following the Civil War. In the low white-collar category, however, cricket club members were representative of Philadelphia society (25.4 per cent against 22.8 for Native-White Americans (NWA) in 1860; 24.0 against 25.0 for Native-White Americans (NWA) in 1870). But the disparity between the club members and NWA returns in the blue-collar brackets. In 1860 less than 18 per cent of a small sample were blue-collar workers compared with 58.6 per cent of the NWA. This differential increased enormously in 1870 as less than 2 per cent of the club members worked in blue-collar occupations compared with 60.4 per cent of the native white Americans.

In addition to holding a disproportionate share of the better positions within Philadelphia's occupational structure, cricket club members held disproportionate shares of real estate and personal property when compared with native-white Americans. Table 10.3 shows that in 1860 14.3 per cent of the NWA owned real estate with an average value of $ 11,008. There were 36.7 per cent holding personal property valued at $2,804 per holder. A smaller percentage of club members owned property – 10.9 per cent held real estate and 18.4 per cent had personal property. Their average real estate holding, though not statistically significant, was $4,000 greater than that of the NWA while their mean personal property more than doubled the amount held by the NWA. The 1870 differentials favouring the club members were even greater. For real estate 15.1 per cent of the club members owned nearly twice as much as the 14.4 per cent of the NWA who held real property. The difference between the two groups was even more pronounced for the personal property variable as the size of the club members' holding was nearly six times greater than that of the NWA. The difference in means tests for personal property holding in 1860 and 1870 and for real estate in 1870 were significant at the .05 level of confidence.

TABLE 10.3

REAL ESTATE AND PERSONAL PROPERTY HOLDINGS OF PHILADELPHIA
CRICKET CLUB MEMBERS AND NATIVE WHITE AMERICANS FOR 1860
AND 1870

	1860		1870	
	Club member	Native white American	Club member	Native white American
Proportion owning $100 or more of of real estate	10.9 (22)*	14.3 (1756)	15.1 (42)	14.4 (1559)
Median value of real estate	$10,000	–	$19,911	–
Mean value of real estate	$15,073	11,008	28,667	$14,924
t-ratio	.059		2.49**	
Proportion owning $100 or more of personal property	18.4 (37)	36.7 (4507)	20.4 (57)	35.0 (3790)
Median value of personal property	$3,000	–	$13,341	–
Mean value of personal property	$5,881	$2,804	$34,883	$5,942
t-ratio	2.66**		2.67**	
Total sample	201	12,280	279	10,830

* Number of cases in parentheses.
** Significant at the 5 per cent level of confidence.

Just as with property holdings and occupation, cricket club members also tended to follow Philadelphia's upper class in residential patterns. E. Digby Baltzell has shown that 'Proper Philadelphians' had established Rittenhouse Square as the city's most fashionable neighbourhood during the latter half of the nineteenth century.[22] This neighbourhood, located west of the central business district (CBD) in wards 7 and 8, attracted nearly 30.0 per cent of the club members in 1860. This proportion nearly doubled in 1870 as 57.3 per cent of the members called Rittenhouse Square their home. The other neighbourhoods in which cricket club members resided in fairly large numbers were certain sections of the Germantown–Chestnut Hill area in wards 21 and 22. Though not as exclusive as Rittenhouse Square, these neighbourhoods became home for a number of Philadelphia élite who had moved there as the city expanded to the north–west. Table 10.4 indicates that 22.4 per cent and 17.6 per cent of the club members were domiciled there in 1860 and 1870, respectively.

TABLE 10.4

RESIDENCE AND WORKPLACE OF PHILADELPHIA CRICKET CLUB
MEMBERS IN 1860 AND 1870

	1860				1870			
		Residence		*Workplace*		*Residence*		*Workplace*
Wards	*No.*	*Percent.*	*No.*	*Percent.*	*No.*	*Percent.*	*No.*	*Percent.*
CBD	15	07.5	90	39.8	12	04.3	153	54.8
Rittenhouse Square	60	29.9	6	03.0	160	57.3	27	09.7
Other adjacent	30	14.9	9	04.5	25	08.9	10	03.6
Germantown area	45	22.4	15	07.4	49	17.6	6	02.1
Other outlying	33	16.4	7	03.5	32	11.5	3	01.2
Unknown	18	08.9	84	41.8	1	00.4	80	28.7
Total	201	100.0	201	100.0	279	100.0	279	100.0

Cricket club members worked predominantly in the CBD. Eighty or 39.8 per cent of the 1860 sample worked there, and 153 (54.8 per cent) of the members had jobs there in 1870. The dominance of the CBD as the chief workplace becomes more evident when the unknown cases are eliminated; then, the percentages climb to 68.4 for 1860 and 76.5 for 1870. In that the majority of the club members held high white-collar and professional positions, they had the financial resources to live in residential neighbourhoods, such as Rittenhouse Square, Germantown and Chestnut Hill, considerable distances from their workplaces in the CBD. Unlike most nineteenth-century Philadelphia workers who had to walk to work, these club members had the luxury of riding, with the exception of the physicians who often worked at their residences.[23] The increased homogeneity of the 1870 sample in workplace and residence as depicted in Table 10.4 is due chiefly to the influx of businessmen into the cricket clubs. They generally worked in the CBD and lived in the élite neighbourhoods.

V

This socio-economic analysis of cricket club members has revealed that they were homogeneous, particularly with respect to birthplace, occupation, residence, workplace and real and personal property. In those demographic characteristics, they tended to take on upper socio-economic class characteristics. The majority were born in Pennsylvania, worked in white-collar and professional occupations, and lived in élite neighbourhoods. While only a small portion held or reported real or personal property, those who did, differed significantly from Philadelphia's NWAs in three of the four property categories. The men who joined Philadelphia's cricket clubs made it possible for the game to prosper there. They had the resources and leisure to maintain cricket, but only a handful spent their leisure hours on the cricket grounds. Less than one-sixth of the club members in the 1870 sample played the game. The majority of the clubs' members supported the cricket aspirations of a few. Their membership fees and other financial contributions purchased the equipment, facilities, and professional instruction which enabled a handful of cricketers, usually the younger club members, to learn and preserve the game of cricket in Philadelphia.

Cricket first appeared in Philadelphia during the 1830s. English immigrants living in the city's Germantown and Kensington sections played the game in their neighbourhoods. These early participants held blue collar positions in the textile mills located in those areas. This finding concurs with the work of Kirsch and Benning. Both have found English cricket players working in blue-collar occupations in Newark and Trenton.[24] Because cricket was part of their culture and because those particular English immigrants had not yet been assimilated into Philadelphia society, either occupationally of residentially, it was natural for them to pursue activities – recreational or otherwise – to which they were accustomed. Their participation in cricket thus had both ethno-centric and recreational meanings for them. Within this context, American youths were first exposed to cricket.

During the 1830s at Haverford College, and in the next decade at the University of Pennsylvania, the motives for playing cricket were somewhat different. At the latter institution, the chief promoters of the game – Robert Waller and John K. Mitchell – believed it had moral implications. Adhering to the sentiment of the British upper class, they viewed cricket as a means of preparing young men for the rigours of adulthood. While that interpretation may have had a certain degree of validity in Philadelphia, the American youngsters and adults played cricket primarily for recreational purposes. Perhaps this viewpoint has been articulated best by William Rotch Wister, the widely recognized 'Father of Philadelphia Cricket', when he described one of the PCC's most devoted players' motivation for playing. He recalled that J. Warner Johnson, a law book publisher and converted baseball player, 'played cricket for exercise and as a recreation from his business'.[25]

Whatever the motives, cricket could not have been established in Philadelphia without the inspiration and help of the British. English residents not only taught Americans how to play, but they also joined their cousins in friendly competition which aroused much interest in the game among the local inhabitants. Then, too, the great international matches of 1859, 1868, and 1872 exposed Philadelphians to superb English cricketers whose brilliant skills and techniques left a lasting impression. Their superlative play inspired native Americans to work harder at learning the English game.

The chief force contributing to the institutionalization of cricket in Philadelphia, however, was the desire of native Philadelphians to master the game and then pass it on to succeeding generations. They did this through the club structure that emerged in Philadelphia during the 1850s. Although transplanted Englishmen like Robert Waller provided the impetus for the club structure, the Americans – who organized and directed the cricket clubs – put forth the effort to ensure the success of the clubs which, in turn, guaranteed the existence of cricket.

At no time was that more evident than during the Civil War years when the war effort and the advent of baseball severely strained cricket in Philadelphia. The young boys from the YACC saved the game during this stormy period. With the return of peace, the rejuvenated Philadelphia and Germantown Clubs, together with the newly organized Merion Club, firmly established cricket in the Quaker City.

The extent of the American role in the establishment of cricket in Philadelphia was uncovered by the socio-economic analysis of the clubs' memberships. Less than 10 per cent of the club members in 1860 and barely more than one per cent in the 1870 sample were born in the United Kingdom. Conversely, more than 50 per cent of each sample was native-born. With respect to occupation, nearly all of the native-born members of the four major cricket clubs had white-collar positions. Before the Civil War, both blue-collar and white-collar workers played cricket in Philadelphia. After the war, blue-collar involvement virtually disappears. This trends denotes one of two things. Either white-collar workers, once they implanted the game, put some distance between themselves and their blue-collar brethren, or blue-collar workers turned to baseball, a game less demanding from the standpoint of time, equipment and facilities. Though Philadelphia cricket never regained its pre-Civil War status, the game, once institutionalized, remained a significant source of pleasure and entertainment for Philadelphia's upper class throughout the nineteenth and well into the twentieth century.

While the English planted the seeds of cricket in Philadelphia, native residents nurtured the game for more than a generation, making it a permanent institution in the city by 1872. Back in 1843 Robert Waller, the Englishman who did more than any other to establish cricket in America, concluded his post-cricket match banquet toast on an optimistic note. Perceiving the favourable atmosphere for cricket in Philadelphia and seeing the game beginning to take hold, he proclaimed Philadelphia 'the nursery of cricket' which 'has produced players qualified to play in any part of the world, and the noble game has taken so deep a root that it can never be uprooted'.[26] Though Waller was perhaps caught up in wishful thinking at the time of his toast, his words carried a truthful ring in 1872.

NOTES

1. This project was supported by a William Paterson College Research Development Award for 1984–85. The writer expresses gratitude to Dr Henry Williams, Associate Director, Center for Philadelphia Studies, University of Pennsylvania, for his direction and assistance in gathering demographic information from the Philadelphia Social History Project data base.
2. W.R. Wister, *Some Reminiscences of Cricket in Philadelphia Before 1861* (Philadelphia, 1904), p. 142.
3. Ibid., pp. 1, 142; F. Lillywhite, *The English Cricketers' Trip to Canada and the United States* (London, 1860), p. 27.
4. Wister, *Some Reminiscences*, pp. 5–15; J. Wister, *Jones Wister's Remembrances* (Philadelphia, 1920), pp. 113–14.
5. Wister, *Some Reminiscences*, pp. 1–19; J.A. Lester, *A Century of Philadelphia Cricket* (Philadelphia, 1951), pp. 11–15; Barnet Phillips, 'Cricket in the Forties', *Harper's Weekly* XXXVIII, 22 Sept. 1894, 908.
6. Minutes of Meetings of the Philadelphia Cricket Club (hereafter MMPCC), 1854–59 (Historical Society of Pennsylvania, Philadelphia); Wister, *Some Reminiscences*, pp. 19, 22–23; Philadelphia *Public Ledger*, 20 November 1927.
7. *Charter, Constitution and By-Laws of the Germantown Cricket Club* (Philadelphia, 1880), pp. 5–6; George M. Newhall, 'The Cricket Grounds of Germantown and a Plea for the Game', *Papers Read Before the Site and Relic Society of Germantown* (Germantown, 1910), 172–86; Wister, *Some Reminiscences*, pp. 25–31.
8. Newhall, 'Cricket Grounds of Germantown', pp. 172–86; Lester, *Philadelphia Cricket*, p. 23; Wister, *Some Reminiscences*, pp. 25–31.
9. *New York Clipper*, 14 July 1860; Lester, *Philadelphia Cricket*, p. 27.
10. *New York Clipper*, 22 Oct. 1859; Philadelphia *Public Ledger*, 27 November 1927; Wister, *Some Reminiscences*, pp. 109–22; Lillywhite, *The English Cricketers' Trip*, pp. 40, 44.

11. The names of several minor cricket clubs, such as Southwark, Oxford, Hamilton, Aramingo, and Mt. Vernon, reflected a neighbourhood image or perhaps the major street which passed through the neighbourhood.
12. MMPCC, meetings of 14 Jan. 1861, 20 Jan. 1862, 16 April 1863; *New York Clipper* 4 July 1863, 31 Oct. 1863; Walter S. Newhall, *A Memoir* (Philadelphia, 1864), p. 16.
13. Newhall, 'Cricket Grounds of Germantown', 192; Charles Blancke, 'Cricket in America', *Harper's Weekly* XXXV, 26 Sept. 1891, 732.
14. G.B. Kirsch, 'The Emergence of Baseball' (manuscript on loan from author), pp. 5, 14–15, 20.
15. MMPCC, meetings of 5 Jan. 1865, 8 Jan. 1866.
16. *New York Clipper*, 1 July 1865, 30 Sept. 1865.
17. *Germantown Cricket Club, Roll of Members and Constitution* (Philadelphia, 1867), pp. 11–17; MMPCC, meetings of 14 Jan. 1867, 17 Jan. 1868, 16 Nov. 1868; *Chronicles of the Merion Cricket Club*, Box No. 1, Historical Society of Pennsylvania, Philadelphia.
18. MMPCC, meeting of 16 Nov. 1868; *Official Handbook of the 1872 International Cricket Fete* (Philadelphia, 1872), pp. 9–12; *Chadwick's American Cricket Manual* (New York, 1873), p. 95; Philadelphia *Public Ledger*, 27 Nov. 1927.
19. *Chadwick's American Cricket Manual*, pp. 92–4.
20. Professor George Kirsch, Department of American Studies, Manhattan College, Riverdale, New York, and I collaborated on collecting demographic information on Philadelphia cricketers in 1860 and 1870.
21. Socio-economic data for members of Philadelphia cricket clubs were obtained from *McElroy's Philadelphia City Directory, 1860*; *Gopsill's Philadelphia City Directory, 1870*; and U.S. Bureau of Census, *Census of Population, Philadelphia, 1860, 1870*.
22. E.D. Baltzell, *The Making of a National Upper Class* (Glencoe, IL, 1958), pp. 174–7.
23. T. Herschberg, *et al.*, 'The 'Journey-to-Work': An Empirical Investigation of Work, Residence and Transportation, Philadelphia, 1850 and 1880', in T. Herschberg (ed.), *Philadelphia; Work, Space, Family and Group Experience in the 19th Century* (Oxford, 1981), pp. 134–8.
24. G.B. Kirsch, 'The Rise of Modern Sports: New Jersey Cricketers, Baseball Players, and Clubs, 1845–1860', *New Jersey History*, 101 (Spring–Summer 1983), 59–67; D. Benning, 'The Emigrant Staffordshire Potters and Their Influence on the Recreative Patterns of Trenton, New Jersey in the Nineteenth Century' (paper presented at the Geographical Perspectives on Sport Conference, University of Brimingham, 7 July 1983), 8–12.
25. Wister, *Some Reminiscences*, p. 143.
26. Ibid., p. 142.

11

South Africa's Black Victorians: Sport and Society in South Africa in the Nineteenth Century

André Odendaal

One of the enduring legacies of British colonialism has been the institutionalization of British sports in the former colonies. This is nowhere better reflected than in South Africa. The British garrison which took control of the Cape in 1806 during the Napoleonic wars, and the thousands of administrators and settlers who followed, brought with them sports like horse racing, hunting and cricket and soon influenced the indigenous people into adopting them. By the end of the nineteenth century sport had become an important social institution and, fitting in with the structure and relations of Empire, South Africa was emerging as a major international sporting nation. Today, as the front-page headlines of the national newspapers regularly indicate, sport occupies a central position in South African life and has become a major issue in the country's domestic and international relations.

The development of sport in South Africa during the nineteenth century was closely linked to colonial politics and reflected in many ways in microcosm the developing South African colonial society and social structures. This could be seen on the institutional level in the organizational structures that developed, in the value systems that became entrenched in sport, in African responses to colonialism, in the role of sport in the process of African class formation, and in the way the development of sport closely followed the pattern of historical development in South Africa. This chapter aims to demonstrate the close connections between sport, politics and social class in South Africa during the first century of British colonial rule by means of a case study outlining the as yet largely unknown early history of African cricket in South Africa.[1]

Not only has the history of black sport been largely undocumented, but, fuelled by the harsh system of racial inequality, segregation and discrimination in South Africa, the myth has arisen that blacks have no real sports history. Thus spokesmen of the apartheid government claim that it is only in the last ten or twenty years that Africans have become interested and started to participate in 'Western' sports: 'For centuries they found their recreation in traditional activities, such as hunting and tribal dances. It was the white nation with its European background and tradition which participated in the recognized sports'.[2]

The unequal development of sport among blacks and whites in South Africa is therefore ascribed 'to some natural differences in the psycho-physiological character of black people' rather than to South Africa's discriminatory racial policies and structures. These myths need to be countered because not only have black people a long, indeed remarkable, sporting history but the development of South African sport has always been closely influenced by wider political and economic factors.

I

The first Africans to be subjected to British rule and become influenced by British values and customs were the Xhosa-speaking people living in the present-day Eastern Cape. For centuries they had lived as pastoral farmers in scattered communities, moving around constantly in search of better grazing. In the second half of the eighteenth century the so-called *trekboers*, the advance guard of white settlement, moving away from the Cape peninsula into the interior, for the same reason, entered their traditional grazing areas. For several decades after the frontier opened the whites and the various Xhosa groups (as well as the indigenous Khoikhoi and San, whose societies were in the process of disintegration) jostled each other for control, but no one group succeeded in asserting its undisputed authority. The balance of power swung in favour of the whites when metropolitan, imperialist Britain took over the Cape. Troops were sent into the frontier to drive the Xhosa out and in 1820 some four thousand British immigrants were settled in the area to act as a buffer between the Xhosa and colony. The assertion of British control did not end there; in the following decades numerous, previously independent, chiefdoms were conquered and incorporated into an expanding Cape colony. By the 1880s hundreds of thousands of Africans had become British subjects.

British rule had a disruptive effect on the conquered societies. A European system of administration was imposed over them and agents of imperialism such as missionaries, teachers, traders and farmers moved into the African territories bringing the indigenous people into contact with alien European ideas and institutions. The missionaries, for example, set up schools and encouraged the people to forgo their 'uncivilized' customs and instead to undergo a basic Western education and to learn Christian doctrine in combination with British cultural values. People were encouraged to wear European clothes, build square houses, give up polygamy and so on. As a result traditional relations and authorities were undermined and new forms of African consciousness and response emerged. These were conspicuously reflected in the emergence of a market- (rather than subsistence-) oriented peasantry and a new class of literate, missionary-educated 'school' people. These people developed into a new distinct, self-conscious élite class which began to grope for involvement in the new economic and social order and to demand political rights for blacks in the Cape political system in line with Christian egalitarian and British liberal political values.

The political system of the Cape provided an outlet for these aspirations as the constitution promulgated in 1853 made no colour distinction. A qualified, non-racial franchise was instituted. This dispensation was based on the prevailing hegemonic and mid-Victorian liberal ideology and the practicalities of free trade imperialism, which emphasized the virtues of free wage labour, secure individual property rights based on a free market and a system of representation. It came to be regarded by Africans as a model system for the colonies, particularly as it contrasted starkly with the other South African colonies, both British and Afrikaner, where there was little pretence of social equality and blacks had virtually no political rights. The African peasantry and the aspiring petty bourgeoisie – teachers, ministers, law agents, clerks, interpreters, storemen, transport riders, blacksmiths, telegraph operators and printers – entertained high hopes that they would eventually be assimilated fully into the evolving Cape society.

Growing in numbers, confidence and assertiveness, the élite had developed into a distinct well-established stratum of Cape society by the 1880s. This could be seen in the process of political mobilization that occurred in that decade. The first modern political organizations were formed, an independent African newspaper was started and around ten thousand Africans registered as voters, enabling them, in certain Eastern Cape constituencies, to return candidates of their choice and, in others, to hold the balance of power. This unique group of enfranchised blacks came to occupy a special position in Cape and South African politics. The political developments were paralleled in other spheres of life as well. Numerous church, temperance, mutual aid, farmers', teachers', cultural and other associations emerged concurrently with the political groups. Politics was only one of a whole range of day-to-day activities in the wider social milieu in which people were responding to new opportunities, and opening up the way for the future.[3]

II

Sport was an integral part of this whole process of assimilation and mobilization. It was one of the many aspects of British culture that the new élite enthusiastically adopted in pursuit of their assimilationist goals. British games, particularly cricket, which the Victorians regarded as embodying 'a perfect system of ethics and morals',[4] were taken almost as seriously as the Bible, the alphabet and the Magna Carta. They were eventually to supersede traditional, pre-colonial forms of recreation in popularity.

Africans were introduced to Western sports both on a formal and informal level, in a way correlative to their other activities. Informally, the Xhosa were interested spectators at the cricket matches and horse races that came to be staged in the new frontier towns that were springing up in the conquered African territory from the onset of such events in the 1850s. A report of a race meeting in King Williams Town being 'enlivened' on the fourth day by ' "Kaffir races" on horseback and on foot', gives a good indication of the informal interaction that was beginning to take place.[5] In a similar vein, an early pioneer in his journal gives details of a meeting in 1862 with a dishevelled-looking farmer on a lonely outpost, three days' ride from the nearest town, who had been 'amusing himself by playing cricket with Kaffirs'.[6]

On a more formal level the emerging class of missionary-educated people were introduced to modern sports at mission schools such as Lovedale, Healdtown and Zonnebloem. These schools, often racially mixed, provided an excellent education based on the English model for thousands of African pupils and fostered the assimilationist ideal. Attendance rose from 2,827 pupils in 1865 to 15,568 pupils in about 700 schools in 1885.[7] Recreation was a matter of supreme importance at these institutions as many of the amusements of tribal Africans were deemed 'incompatible with Christian purity of life' and had to be abandoned by those embracing the new religious ideas of the missionaries. Provision was, therefore, made for the 'profitable employment of leisure'.[8] Drill became a regular feature on timetables and sports like cricket and football were introduced. At a mission in Natal, an observer noted in 1857 that the skill of African boys in flinging assegais gave them an advantage over white boys at cricket: 'they rarely fail to strike down the wicket from a distance'.[9] At

Zonnebloem College, which was started by the Cape Governor, Sir George Grey, with the aim of acculturating or 'civilizing' the sons of chiefs, the college records describe the enthusiasm shown by the pupils for cricket after its inception there in 1861. Within three years the school was fielding two sides and playing matches against other schools.[10] In 1910 the mayor of Cape Town 'remembered a time when the College had the best cricket team in the whole Peninsula'.[11] That this mission education made no small impact on the students is well illustrated in the case of Nathaniel Umhalla, the son of the last independent Ndlambe chief, who became one of the most respected mission-educated figures in the eastern Cape. In addition to undergoing a British education and becoming literate, he acquired an English name, adopted Christianity as his religion, visited Britain, took up a job in the civil service, became actively involved in colonial politics and, naturally, retained a lifelong interest in cricket. In 1870 he was playing with two other ex-pupils in the mixed St Mark's mission side against the white Queenstown Club and 15 years later was still prominent as a player and administrator in King Williams Town. Also instructive of the origins of sport among Africans and the interrelationship between religion, education, culture and sport was a report in a missionary newspaper of festivities in 1870 to celebrate the founding of one of the earliest African Sunday School Unions. After a day and a half of church services and festivities the nearly 700 young people involved 'broke up into parties for various sports, among which the English game of cricket attracted many of the elder boys and young men'.[12] Typically, on the Queen's birthday in 1877, all the pupils at Lovedale had a day of sports in the fields.[13] Through their new education, as well as economic and religious activities, Africans were adapting to Western ways and beginning to internalize many Western values.

Sport as we know it was still in its infancy in South Africa by the 1870s, though it had by then spread to all parts of South Africa. Only a few clubs had been established in larger centres like Cape Town, Pietermaritzburg and Port Elizabeth; among them the first African cricket club started in the last-mentioned town in 1869.[14] No regional or national associations had been formed and there were no official leagues or competitions. However, from 1775 to 1885, coinciding with the rise of sport as a mass leisure activity in the new post-Industrial Revolution environments in Britain, a number of new sports were introduced to South Africa and sport became institutionalized there as well. That decade saw the formation of the first rugby, football, athletics, cycling, horse racing (jockey), golf and tennis clubs and the inauguration of regular competitions. Then from the late 1880s

13. The making of South Africa's black Victorians – Lovedale pupil teachers at drill (J. Wells, *Stewart of Lovedale*, London 1898)

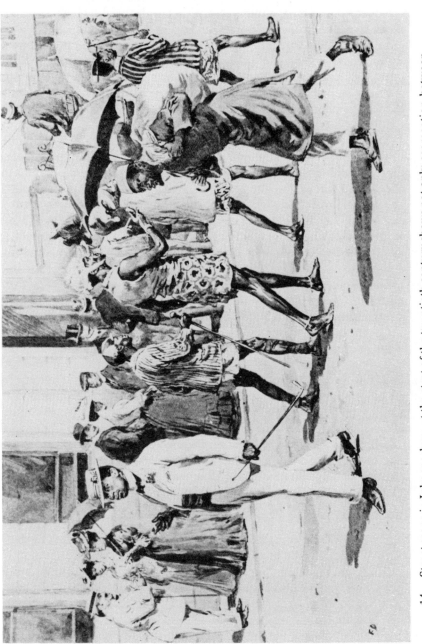

14. Street scene in Johannesburg at the start of the twentieth century shows not only segregation between black and white but also emerging class differences among blacks (*The Graphic*, 17 March 1906)

national associations started coming into existence to place sport on an organized footing.[15] The main reason for this growth was the discovery of the richest mineral deposits in the world which attracted thousands of European fortune-seekers into the interior, stimulated industrialization and urbanization and led to large-scale imperial expansion which was to culminate in the unification of the various territories into a single British colony – South Africa – in 1910. South Africa's industrial revolution set the stage for the rise of sport as a modern phenomenon with mass appeal in much the same way as the British Industrial Revolution had done.

The African élite in the Eastern Cape were not slow in responding to these impulses. By the mid-1880s, following the white precedent, there were thriving African cricket clubs and regular competitions in almost all areas in the region.[16] The first African-controlled newspaper, *Imvo Zabantsundu* (Native Opinion), which was started in 1884 and is still in existence today, abounded with reports. These were printed under the headings of 'Ibala labadlali' (sports reports), and by 1887 the newspaper had a 'sporting editor'. In the advertising columns the big Dyer and Dyer merchant house placed advertisements directed specifically at African cricketers and clubs.[17] The *Port Elizabeth Telegraph* was not exaggerating when it observed that cricket 'seems quite to hit the Kaffir fancy'.[18]

Competition was placed on a more co-ordinated footing in 1884 when teams from the main Eastern Cape centres – East London, King Williams Town, Queenstown, Grahamstown and Port Elizabeth – took part in the first of several inter-town tournaments. These were based on similar inter-town tournaments for the Champion Bat among the best white teams in South Africa.[19] Smaller competitions involving African sides from other centres in the Eastern Cape were also held. In addition, without any South African precedent to go by, plans were set in motion to send a combined side chosen from the best players at the inter-town tournament on a tour of England.[20] Unfortunately these plans went unrealized. It was to be a decade before the first (all-white) touring team left South African shores. Enthusiasm for the game continued to grow in the 1890s; just one indication of this was the 15 fixtures played by the Frontier Club of King Williams Town in a season in mid-decade.[21]

While cricket was by far the most popular sport, the aspiring black petty bourgeoisie also took, to a lesser extent, to sports like tennis, croquet, football and rugby. By the end of the 1880s there were tennis and croquet clubs in several towns. These two sports attracted women competitors as well. At the helm of the Port Elizabeth ladies' croquet club, formed in 1884, were the wives of several prominent church and political figures – no doubt Victorian ladies in the proper sense of the word.[22] Football and rugby, which became the next most popular sport in the Eastern Cape after cricket, took root in the 1890s. More unusual was the report, also in the 1890s, of the horse races held by the Queenstown Africans on the local showgrounds. The prosperous Meshach Pelem, a prominent politician, won the one-mile pony plate with his 'Little Wonder', and gained a place in another race with another horse.[23]

The King Williams Town cricketers gave an indication of the proficiency of black sportsmen when, just after winning the first inter-town tournament, they challenged and beat the local white town team which had recently taken part in the corresponding white inter-town tournament.[24] At the same time the *Imvo* recorded a victory by the Port Elizabeth African team over the white Cradock side.[25] These cases were by no means exceptions. Black teams regularly beat white sides on the special occasions, usually public holidays, that they played together.[26]

III

The development of sport in the 1880s was an integral part of the wider process of African political mobilization that was occurring in the Eastern Cape at the same time. A whole new framework of interrelated activity based on western models was emerging as people organized at every level. The case of John Tengo Jabavu, the most important black spokesman of his age, illustrates this strikingly. In addition to buying a farm, starting a newspaper and rising to prominence as a political figure, he was a Wesleyan church steward and Templar and also became chairman of two of the cricket and lawn tennis clubs that were formed in King Williams Town. Moreover, he presented the Jabavu Cup for inter-town competition in cricket. Jabavu was not the exception. The early political leaders were almost invariably also leaders and members of the first sports clubs. Rank here added to their social status.[27] Conversely, sport also intruded directly into political life. In 1884, for example, a paper extolling the benefits of sport was read before the pioneering Native Educational Association.[28] A few years later, coloured people in Port Elizabeth were excluded from membership of a new black trading co-operative, the African and American Working Men's Union, because they regarded themselves as too 'high' to play sport with Africans.

Sport, particularly cricket, served an explicitly political function for the black élite. They were intent on using it as an instrument of 'improvement' and assimilation. By enthusiastically playing the most gentlemanly and Victorian of games, they intended to demonstrate their ability to adopt and assimilate European culture and behave like gentlemen – and by extension to show their fitness to be accepted as full citizens in Cape society. Through sport they could pay homage to the ideas of 'civilization', 'progress', 'Christianity' and 'Empire' that were so precious to the Victorians, and call for imperial concepts of 'fair play' to be respected; through sport they could assert their own self-conscious class position.[29] Given the realities of life at the Cape it is not surprising that they held these idealized values dear, and that despite the obvious contradictions they glorified things 'British' (the ideal) as against things 'colonial' (the reality). It followed that when the first English cricket side toured South Africa in 1888/89 the black sportsmen – in an obvious political commentary – cheered them on against the local white sides. In the report on the match in King Williams Town, the *Imvo* noted, 'It is singular that the sympathies of the Native spectators were with the English.'[30]

The determination of the new élite to be accepted as fellow Victorians and citizens by whites in colonial society was well reflected when the King Williams Town African team beat their white counterparts. Commenting on the win, despite a lack of experience and facilities, John Tengo Jabavu declared: 'It is enough to say that the contest shows that the Native is a rough diamond that needs to be polished to exhibit the same qualities that are to be found in the civilised being, and that he is not to be dismissed as a mere "schepsel", as it has been the habit of the pioneers to do so hereto'.[31] Such cricket matches, moreover, were 'calculated to make the Europeans and Natives have more mutual trust and confidence than all the coercive and repressive legislation in the world'.[32] On the proposed tour of England, he said it would 'also afford our friends there the opportunity of realising the tone that European civilisation gives to the society of Africans'.[33]

Just as important as the actual games, and just as instructive of the new forms of socialization based on colonial models and the class position of the relatively prosperous educated élite, were the social activities connected with them. Sporting contests on public holidays such as Empire Day and Christmas were almost inevitably followed by social functions. As Willan has noted in his study of Kimberley, these were clearly derived from the Western model, and differed little in this respect from those that catered for white bourgeois society in that town. Often functions were held in the town hall, with the mayor or other dignitaries in attendance. Willan describes one typical such occasion: a splendid dinner was put before the guests, after which a programme of musical entertainment and speeches followed. The latter was started off with a toast to the Queen and ended with a rendering of 'God Save the Queen'. Musical items included 'What can the matter be' and 'We shall meet again'. Finally, the proceedings were brought to a close with a hymn and a benediction.[34] In a similar vein, the King Williams Town cricketers on one occasion held a special farewell reception at a hotel owned by one of its wealthiest members – the first such African enterprise in the colony – before travelling down to Port Elizabeth by sea in the new Dunvegan Castle liner to meet their old rivals. Fellow passengers included Prime Minister Rhodes and his entourage on their way back from Rhodesia.[35]

One could not wish for a greater reflection of the changing value systems and experiences accompanying colonization than these adventures. They reflected not only the strong desire of the élite to be assimilated into colonial life, but also the opening up of class cleavages among Africans. In adopting British cultural values and seeking upward mobility in Cape society, the aspiring black petty bourgeoisie often distanced itself from the mass of Africans who remained within the traditional framework or were becoming proletarianized. Thus, when a Bill was introduced into the Cape Parliament in 1891 to prohibit certain 'obscene' tribal amusements, Jabavu's newspaper encouraged Parliament, in the language of the rulers, 'not only to pass stringent measures to suppress the (traditional initiation) dances, but to render it unsafe for boys submitting to the barbarous rite to appear in places of public resort, where their presence is calculated to place a bad example to young men endeavouring to cultivate good morals'.[36]

Clearly sport and the related social activities were providing the new élite with a social training ground for participation in the new society, and in typical Victorian fashion it provided both a personal and political lesson for them. A member of the African Political Association (APO), the earliest coloured political organization, speaking on the topic of a sound mind in a sound body, emphasized just how closely sport and politics were linked:

> . . . great lessons can be learned . . . on the cricket and football fields – two forms of sport of which our people are passionately fond. No one who is not punctual, patient, accurate and vigilant, can ever expect to become a consistently good batsman. Both batsmen and spectators know that; and yet do we carry those moral lessons into our private or public life? Patient, of course, we are: but are we punctual and vigilant? Often, a Chairman of a Branch of the APO is half-an-hour late, or the Secretary has not

his minutes written up, with the result that the meeting is out of temper for the rest of the evening and adjourns half-an-hour late . . . Again, are we as watchful of our public welfare as the batsman is of every ball – even those which the umpire declares to be 'wide'? If we were, much of our present trouble would have been forestalled.

Now turn to the football field, and watch a match, say between the (white) Hamiltons and Stellenbosch. The forwards present an invulnerable front. In the scrum they pack into one inseparable whole, and press forward with a regular rhythmic movement and a steady earnestness of purpose. They are not easily broken up. There is *perfect union*. Again the backs, on whom the eyes of the spectators are concentrated, obtain the ball. They rush down the field, and secure a try. Why? Because there has been a thorough combination. The passing was accurate and well-timed; and finally, there was a *complete subordination of self*.

Our young men are good cricketers, but poor footballers. The reason for this is apparent. In cricket, individual excellence often wins the game: while the result of a football match depends much more upon union and combination and subordination of self, than the strength and agility of any particular player.

And, as on the football field, so in the battle of life. We lack union: we refuse to combine; and self is not sufficiently suppressed in the interest of the people as a whole. These failures are not, fortunately, inherent in us . . . There is hope that in future the moral lessons to be learned on the cricket and football fields will be carried into public life for the benefit of the people as a whole.[37]

The assimilationist ideals of the black élite in the Cape, however, were to become increasingly frustrated both on the field of sport and in the wider political arena. Though white paternalism allowed for the odd sporting encounter with blacks, social segregation was the norm, and whites had no intention of relaxing the barriers. Some mixing may have occurred in mission teams but clubs remained strictly segregated. This obtained until the mid-twentieth century in other British colonies such as Kenya, Nigeria and Ghana.[38]

In the British African and Asian possessions the club served as a symbol, not only of social status, but also of political domination, as Jan Morris has pointed out. 'There it was developed as an enclave of power and privilege in an alien setting, its members patently different from the un-admitted millions not only in colour and status, but also in place. More than anywhere else, it was the place where the imperialists celebrated their Britishness, authority and imperial lifestyle.'[39] The social exclusivity of the Victorians in the colonies went hand in hand with the most prejudiced feelings of cultural superiority – even towards other Europeans.

'We are, in fact,' wrote A.G. Leonard comparing Briton and Boer, 'the running water of a mighty river of humanity flowing unchecked and irresistible, while they are stagnant slime-covered moss.'[40] Blacks were even further down the scale. Observing the slaughter of wave after wave of Muslim fighters by the recently invented machine gun at the Battle of Omdurman, the young Winston Churchill said: 'These extraordinary foreign figures . . . march up one by one from the darkness of barbarism to the footlights of civilisation . . . and their conquerors taking their possessions, forget even their names. Nor will history record such trash'.[41] Echoing these Victorian stereotypes and referring directly to sport, John Buchan declared:

> On the lowest interpretation of the word 'sport', the high qualities of courage, honour, and self-control are part of the essential equipment, and the mode in which such qualities appear is a reflex of the idiosyncrasies of national character . . .

> It is worth while considering the Boer in sport, for it is there he is seen at his worst. Without tradition of fair play, soured and harassed by want and disaster, his sport became a matter of commerce, and he held no device unworthy . . . (The Boers) are not a sporting race . . .[42]

Yet, despite this snobbery and the long and bitter racial rivalry between English and Afrikaner, and reflecting more the power relations in colonial society and the need for imperial assimilation, the two groups became increasingly integrated on the sports field. By the early twentieth century Afrikaners were becoming influential in South African sport, and old social barriers were falling. Blacks, however, were affected in exactly the opposite way. The paternalistic treatment they sometimes received initially in social and political life grew into a rigid system of segregation in the twentieth century. Highlighted here were the contradictions in the ideologies of sport and imperialism. On the one hand they epitomized 'fair play' and the ideals of the black élite; on the other they entrenched white racial and class attitudes and domination.

An event which underlined the limits to black advancement in sport in the Cape occurred in 1894 when Krom Hendricks, a coloured cricketer said by English Test players George Rowe and Bonnor Middleton to be one of the fastest bowlers they had encountered, was first included in the final squad of 15 players from which the side for the first South African tour to England was to be selected, but later omitted as a result of 'the greatest pressure by those in high authority in the Cape Colony'.[43] Other examples of discrimination also occurred. A few weeks after the above mentioned match, in which the King Williams Town Africans beat the white side, the Town Council barred Africans from the pavilion they had recently used.[44] Local authorities also made life difficult for black sportsmen wishing to use local facilities.[45] Structural restraints such as these were increasingly to frustrate the developmental ambitions of black sportsmen.

The hostility on the part of many white colonists to black middle-class advancement and leisure activities was well expressed by the resident magistrate of Adelaide who recommended, in 1908, that a law should be passed to force Africans to 'understand that work is no crime'. He said that the educated African's sole idea is to copy the European: 'with a white cricket coat and trousers, he is great at tea-meetings, cricket and tennis parties, but he thinks that to do an honest day's work is far beneath his requirements'.[46] The message was clear: blacks should not aspire to social equality; their proper role was to be a labouring class.

IV

The attack on the African élite occurred on a much wider level than merely the sporting one. From the 1880s South African policies became increasingly based on the broad principles of restricting social integration, ensuring an adequate supply of disciplined and inexpensive black labour for white enterprise, and hampering African access to skills, organization and land.[47] In 1887, for example, the Cape Parliament passed the Voters' Registration Act which raised the franchise qualifications and excluded Africans living on tribal tenure from the vote. Just under ten thousand African voters launched a large-scale agitation to have the legislation withdrawn, but without success. Thousands were struck off the roll. The Voters' Registration Act (known as *Tung'umlomo*, the muzzling or sewing up of the mouth) was the first in a long series of measures intended to exclude Africans from the country's political process. In 1892, after the number of African voters had again risen to the old levels, the Franchise and Ballot Act again raised the franchise qualifications. In 1894 Cecil Rhodes piloted through Parliament the Glen Grey Act which aimed to push Africans off the land into the labour market, and excluded those people who were subject to its provisions from the vote. In 1899 the Anglo-Boer War broke out and Africans supported Britain in the hope that the Cape franchise system would be extended to the Orange Free State and Transvaal in the event of a British victory. Their hopes were dashed when the British agreed during the peace settlement to maintain a political status quo which denied blacks any political rights.

The war brought the whole of South Africa under British rule for the first time, and set in motion the process of unification which led to the birth of the Union of South Africa in 1910. Africans protested against the clauses of the peace treaty and demanded the extension of the franchise to Africans in all the colonies in any Union. African political organizations and newspapers similar to those in the Cape had, by now, also been started in Natal, the Transvaal, Orange River Colony and Basutoland. But again the trend was towards the restriction rather than the extension of African political rights in the South African colonies. The constitution of the new Union took away the right of Cape African voters to become Members of Parliament and left blacks in the other provinces without any franchise rights whatsoever. Africans protested strongly against the terms of Union, but to no avail. Even a special delegation of black politicians to London could not persuade the British government to insist on amendments to the colour bar clauses before ratifying the constitution of the new South African state. Britain's economic and strategic interests far outweighed any humanitarian concern it may have felt for Africans, so the imperial government countenanced a constitutional system in South Africa which contradicted its own political democratic ideals.

The explanation for the hardline direction in South African racial policy and the failure of Britain to protect African rights lay mainly in the political economy. The rapid economic development caused by the mineral discoveries in the last quarter of the nineteenth century was to transform and fundamentally re-shape South African life. Ideologies and state structures changed in line with the needs of the developing capitalist economy. Among other things 'Cape liberalism' was undermined – a step which had serious consequences for the enfranchised black élite in that colony. The demand from industrial capital for a huge supply of labour for the diamond and gold mining centres, accompanied by a growing need for labour by a Cape economy stimulated by the mining revolution, led to a change in the material base of Cape liberalism and eventually to the acceptance of a segregationist ideology. Whereas the nature of the Cape economy before had been to encourage African development and enfranchisement, thereafter African advancement clashed with the needs of a growing capitalist economy. A free independent peasantry with access to the political machinery was not conducive to the formation of a cheap, controlled labour force, so the trend in the following years was for the whites to place statutory restrictions on African economic advancement and to restrict African political rights.[48] The failure of political liberalism in South Africa – symbolized by Union in 1910 – also marked the failure of the moderate, constitutional assimilationist strategy of the black élite, which had been tied to the fortunes of liberalism in the Cape. Instead of gaining for themselves an extended role in a non-racial political system, they were to come under increasing pressure in a system which institutionalized racial discrimination and were eventually to be deprived of the franchise rights they had enjoyed for nearly a century. Having deeply integrated South Africa into the Western capitalist economy, Britain had no intention of forcing political changes that would upset the relations of capital, even if it meant legitimizing institutionalized racism in that country.

V

Besides fundamentally affecting South African life in general, the mineral discoveries directly influenced the subsequent spread and nature of black sport in South Africa. Accompanying the many enthusiasts attracted by diamonds and gold, the new games rapidly took root in the interior; national associations were formed in line with the wider process of economic and political integration that was occurring; and the massively rich and powerful mining industry, which soon became the backbone of the economy, started to an ever-increasing degree to control the development of black sport. In the industrialized, urbanized, rigidly race-stratified South Africa of the twentieth century black middle-class pursuits like cricket were to go into decline to be supplanted by working-class mass sports – soccer, boxing and athletics – and a culture which more accurately reflected the living conditions and status of blacks in South African society.

Among the hundreds of thousands of people who converged on the new mining centres in the late nineteenth century, or started working on the developing infrastructure unfolding magnetically northwards towards Kimberley and the Witwatersrand, were many members of the eastern Cape élite. With their unique educational qualifications they generally occupied the most sought after and best paid jobs available to blacks. They naturally assumed too a position of social dominance and leadership among the increasingly cosmopolitan communities in the new industrial centres where members of many different chiefdoms were conglomerating. No less than on the mission stations of the eastern Cape the ideas of 'progress' and 'civilization' remained important to the élite, who took the lead in starting new choral, church, mutual improvement and sporting associations.[49] Duncan Makohliso, prominent in eastern Cape politics, for example, started a tennis club in Bloemfontein while working on the construction of the railway line to the north. Also indicative of the eastern Cape influence were the contests inaugurated between teams from that region and rapidly developing centres like Kimberley, Bloemfontein and Johannesburg from the late 1880s onwards.

After the discovery of diamonds Kimberley soon became one of the most important sporting centres in the country for both black and white. By 1888 (the year in which Griqualand West became the first holders of the Currie Cup competition for white cricketers) cricket organization among Africans there had developed to the extent that a team was sent to play against Port Elizabeth for the first time. Two years later Kimberley participated in the 1890 inter-town tournament against King Williams Town, Grahamstown and Port Elizabeth. There were two clubs for the Kimberley Africans, each of which ran several teams. Fixtures between them were big social occasions. Reflecting the cosmopolitan urban environment they also played against local Indian, Malay and coloured teams. These sides too were composed of migrants, particularly from Cape Town, and they in turn were in touch with their counterparts elsewhere. In 1890 Kimberley hosted a tournament of Malay sides, following a similar tournament the year before in Cape Town. In order to regulate the contests between the various local clubs a Griqualand West Coloured Cricket Union representing all black cricketers in the area was formed in 1892. Two years later a similar regional rugby board was formed with cricket personalities like Isaiah Bud Mbelle, an Eastern Cape African, prominent.[50]

Soon the Kimberley-based sportsmen were initiating moves to co-ordinate competition on a national level in the same way as white sportsmen. In 1897 the South African Coloured Rugby Board was formed in the city, after Bud Mbelle, who became the secretary, had persuaded Cecil Rhodes, the mining magnate and arch-imperialist, to donate a trophy, the Rhodes Cup, for inter-provincial competition along the lines of the newly-instituted Currie Cup.[51] Following this up Bud Mbelle, on behalf of the Griqualand West Coloured Cricket Union, approached Sir David Harris of de Beers diamond company for a similar trophy for cricket.[52] The Union duly received a silver cup worth one hundred guineas, called the Barnato Memorial Trophy, in honour of another mining magnate. The new Barnato Trophy was contested for the first time at a tournament in Port Elizabeth in December 1898.[53] But for some reason cricket did not follow rugby in forming a national body at this time. And another Barnato tournament was not held for several years. This was due to the outbreak, in October 1899, of the Anglo-Boer War, which lasted for almost three years. However, soon after the war, a South African Coloured Cricket Board (SACCB) was formed and the Barnato tournaments were resumed. Black cricketers of all shades were catered for under this arrangement for the next two decades.[54]

The initiatives in the formation of clubs, regional associations and competitions had come from the black élite themselves but, as the presentation of the Rhodes and Barnato trophies indicated, the mining industry began to play an increasing role in the development of black sport. It came to see sport as an important means of social control, not only helping to accommodate and channel the social aspirations and needs of the small petty bourgeois élite, but also to ensure discipline and productivity among the mass of non-literate, menial workers. Under the compound system, hundreds of thousands of black male workers came to be housed in harsh, strictly controlled conditions. To deflect their attention away from the beer-drinking, prostitution and faction fighting (and later also political discontent) that were common in the harsh mining environments, management initiated organized recreation. The rationale behind this was clearly expressed by a Rhodesian official:

> For a moment let us consider what it was that made the (pre-war) British proletariat contented, although working in many cases, in circumstances which were scarcely more conducive to a sustained interest in their actual labours than are those in which the mine boys work here. It was largely sport – or what the workman considered sport. For example, the hands old and young in every community were enthusiastic 'supporters' of some local football team whose Saturday afternoon matches furnished a topic of interest for the remainder of the week. Here the labourers' principal recreations are connected with beer and women, leading frequently to the Police Court and the risk of being smitten with one or other of the venereal diseases which are so insidiously sapping the strength of the native population. Those who employ and those who control native mine labour should, for a double reason, try to influence the native to change in this respect. Sporting enthusiasm is not the ideal substitute for present conditions but it would be a step forward, and one, I am

sure, not difficult to bring about. The native is intensely imitative, often vain, and always clannish, and all these are qualities which would further 'sport' – a parochial spirit of sport if you like – but one which would forego ties of interest and esprit de corps between the labourer and his work-place. A patch of ground, a set of goal-posts and a football would not figure largely in the expenditure of a big mine.[55]

In the twentieth century, recreation facilities became a common feature on the mines, with management, in addition, instrumental in initiating and sponsoring many competitions on a wider community level. In this way the basis of sport among black people was widened and ultimately black sports became working-class in nature. The petty bourgeoisie, which often occupied an intermediary position between management and the mass of workers on the mines, clung to its shrinking base and upper middle-class games like cricket and tennis; but they no longer reflected the material context and became more élitist in nature, going into decline instead of expanding, and exacerbating class differences. The mines and the system of social control operating there revealed much about the underlying processes at work in South African society.

VI

By the beginning of the twentieth century, then, British sports were being played by the indigenous people throughout South Africa, even as far afield as Rhodesia (now Zimbabwe) where the South African settlers who had accompanied the conquering British pioneer column in the 1890s were taking the lead.[56] The first Barnato cricket tournament reflected the increasing spread of the game. Teams from Kimberley (Griqualand West), Cape Town (Western Province), East London, King Williams Town (Southern Border), Port Elizabeth and Queenstown participated, and a side from Johannesburg was at one stage expected to take part but did not. The 'Moslems' from Western Province won the trophy after recovering in one of their matches to beat Southern Border, who had made them follow-on after a first innings collapse. In its reports on the tournament the *Imvo Zabantsundu* newspaper confirmed that the Cape men were by far the most accomplished players.[57]

Probably closer to white society than anywhere else in South Africa in this cosmopolitan and until recently, racially mixed city, the so-called Malay and coloured people of Cape Town became knowledgeable and proficient sports followers at an early stage.

Hattersley, for example, writes of gay scenes at the horse races as early as the 1820s where '. . . Malays and Negroes mingled with whites, all crowding and elbowing, eager to get a sight of the momentous event'.[58] The standard of cricket was such that a Malay team was actually given a fixture against W.W. Read's 1891/92 English touring side. They lost by ten wickets, but Krom Hendricks took four wickets for 50 runs in 25 overs and L. Samoodien hit 55, one of only two South Africans to reach 50 on the tour.[59] In 1894, Hendricks was included as a fast bowler in the final squad of fifteen for the first tour tŏ England by a South African side, but as already indicated, he was later omitted as a result of political pressure. Often black players took part in the practice sessions of local white clubs and visiting teams. In his book on the 1905/06 MCC tour, Sir Pelham Warner described how, at practice, C.J. Nicholls, 'a young Malay with a fast left-hand action hit my middle stump nearly every other ball'.[60] The Cape people were also keen spectators at the famous Newlands cricket ground, starting a long tradition of enthusiastic support for the game there.[61] On an organizational level clubs, unions and leagues, such as the Bailey Shield Competition, had been started by 1900.[62]

Even in the Afrikaner republics in the north – the then Orange Free State and South African Republic (Transvaal) – where a rigid system of social and political segregation existed, blacks were playing the Englishman's game of cricket by the late nineteenth century. In contrast to the Cape, strong deterrent laws forbade blacks in the Transvaal from walking on the same pavements as whites and even from watching white sporting contests[63] but these were not enough to stop them from imitating and adopting the hegemonic cultural values and games. Nor were the sometimes disapproving attitudes of certain continental missionaries, perhaps not as convinced of the beneficial attitudes of British games on Africans as their English counterparts in the south.[64] Reports of cricket in the republics go back to at least 1890. In that year the Potchetstroom Native Cricket Club played against the Kroonstad club in the last mentioned town. Because of its central situation Kroonstad was a popular venue for contests between teams from the two republics. Details also appear in the *Imvo Zabantsundu* newspaper in the 1890s of games between teams from Bloemfontein and Rouxville in the southern Orange Free State and their Cape counterparts in the towns just across the Orange River, as well as between sides from Johannesburg and the Cape. In 1896 the Indian cricketers in Johannesburg established the Transvaal Indian Cricket Union and two years later the Africans formed a similar regional body; the secretary predictably was an eastern Cape man who had joined the influx to the Rand. In 1904 Transvaal became an affiliate of the SACCB and started participating in the Barnato tournaments.[65] In an exception to the rule in the Orange Free State mixed contests sometimes took place at the Baralong enclave of Thaba Nchu between the local team – drawn from a small prosperous, highly politicized land-owning élite[66] – and the neighbouring white cricketers, possibly missionaries and British soldiers who had been granted land in the area after the Anglo-Boer War. Commenting on these matches which the Africans won more than once, a newspaper financed by the local élite commented that whites held themselves socially aloof in order to command respect from blacks but 'the fact is no Natives respect their European neighbours as much as the Baralongs at Thaba Nchu . . . In other parts, where the whites will not play them the coloureds boast that the whites are afraid of them'.[67]

Despite the fact that Natal was a British colony, cricket did not get going there among the indigenous Zulu people in the same way as among the Xhosa in the Cape. The odd cricket reports appear in the early *Inkanyiso lase Natal* (Light of Natal) and *Ilanga lase Natal* (Sun of Natal) newspapers, but it was football that took the fancy in that colony from the start. This was largely due to the influence of the American Board Mission which dominated the Natal mission field. Cricket became, instead, the game of the Indian merchant and trading class with whom Gandhi, who lived in South Africa for 20 years after completing his studies in England, was closely associated.[68] In 1896 they formed the Durban and District Cricket Union. By 1913, when a Natal side played in the Barnato tournament, they had linked up with black cricketers in the rest of the country. The names of the players – Sigamoney, Soobrial, Kaisovaloo, Christopher, Subban, Bhugwan, etc. – confirmed where the interest in cricket was centred in that region.[69]

Thus by the time of Union in 1910 cricket was being played by black people throughout South Africa, the nature of the game in the various areas having been shaped by patterns in South Africa's historical development. In this early period the development of black cricket also closely followed that of white cricket in many respects; from its introduction into schools, to the formation of clubs, to the introduction of inter-town competitions, leagues and provincial competitions through to the formation of a national controlling body. As we have seen, enthusiasm and playing standards in the Cape often matched those of Europeans. The yawning disparity was to develop in the twentieth century as racism, political oppression and economic exploitation intensified and became institutionalized in the modern

South African state, leading to the decline of the black (especially African) petty bourgeoisie, and to the consequent frustration of their middle-class ideals.

NOTES

1. When writing on South Africa one has to contend with the problem of racial terminology. 'African' refers to the country's indigenous Bantu-language-speaking inhabitants. 'Black' is used to describe African, Asian and so-called 'coloured' people collectively; the term has become popular as a positive means of expressing solidarity between all those people not genetically classified as 'white' and subjected to discrimination under South African race laws. Words that now give offence (for example, Kaffir, Native, Boer) appear in the text only in direct quotations.
2. R. Archer and A. Bouillon, *The South African Game : Sport and Racism* (London, 1982), pp. 8–9.
3. For details of the processes of incorporation and political mobilization see A. Odendaal, *Black Protest Politics in South Africa to 1912* (Totowa, NJ, 1984) and A. Odendaal, 'African political mobilization in the Eastern Cape, 1880–1910' (unpublished PhD thesis, University of Cambridge, 1984).
4. K.A.P. Sandiford, 'Cricket and the Victorian society', *Journal of Social History*, Winter 1983, 303.
5. G.S. Hofmeyr, 'King Williams Town and the Xhosa, 1854–1921 (The role of a frontier capital during the high commissionership of Sir George Grey)' (unpublished MA dissertation, University of Cape Town, 1981), pp. 58–9.
6. A.F. Hattersley (ed.), *John Sheddon Dobie South African Journal, 1862–6*, Van Riebeeck Society, first series, No. 26 (Cape Town, 1945), 46.
7. Archer and Bouillon, op. cit., p. 26.
8. R.H.W. Shepherd, *Lovedale South Africa, The Story of a century 1841–1941* (Lovedale, 1940), p. 508. For more details on the missionary institutions see J. Hodgson, 'Zonnebloem and Cape Town, 1858–1870' (paper delivered at Cape Town History Conference, University of Cape Town, 1981).
9. A.F. Hattersley, *An Illustrated Social History of South Africa* (Cape Town, 1969), p. 221.
10. J. Hodgson, 'A History of Zonnebloem College 1858–1870, A Study of Church and Society' (unpublished MA thesis, University of Cape Town, 1975). I am grateful to Janet Hodgson for this and several other references to sport in this period.
11. 'Zonnebloem College', *African Political Association (APO)*, 24 Sept. 1910.
12. *The Kaffir Express*, 1 Dec. 1870, p. 4.
13. Cape Archives, NA 467, Ecclesiastical 1875–1890, J. Buchanan to C. Brownlee, 21 June 1877.
14. 'Iziqingata neziqingata', *Imvo Zabantsundu*. 10 June 1897.
15. Archer and Bouillon, op. cit., pp. 22–4.
16. Paper read by Elijah Makiwane to the United Missionary Conference, *Imvo Zabantsundu*, 19 July 1888.
17. 'Ixesha le bhola, 1889', *Imvo Zabantsundu*, 17 Oct. 1889.
18. Quoted in *Imvo Zabantsundu*, 22 Dec. 1884.
19. For the early history of cricket in South Africa see L. Duffus, M. Owen Smith and A. Odendaal, 'South Africa' in E.W. Swanton (ed.), *Barclays World of Cricket : The Game from A–Z* (London, 1980), pp. 103–11.
20. Editorial notes, *Imvo Zabantsundu*, 3 Nov. 1884. See also 'Notes on current events', *Imvo Zabantsundu*, 4 Oct. 1888 and 'Correspondence', *Imvo Zabantsundu*, 11 Oct. 1888, 18 Oct. 1888 and 20 Dec. 1888.
21. 'Ibala labadlali', *Imvo Zabantsundu*, 2 July 1896.

22. 'Izimiselo ze Kroki', *Isigidimi sama Xhosa*, 16 June 1884.
23. Odendaal, 'African political mobilisation', pp. 297–8.
24. 'Amangesi nabantsundu', *Imvo Zabantsundu*, 2 March 1885.
25. Editorial notes, *Imvo Zabantsundu*, 2 March 1885.
26. See, for example, editorial notes, *Imvo Zabantsundu*, 16 Feb. 1885; 'Ibola e Komani, *Imvo Zabantsundu*, 9 Dec. 1885; 'Ibala laba dlali', *Imvo Zabantsundu*, 21 Dec. 1887.
27. Odendaal, 'African political mobilisation', pp. 299–300. For the development of rugby in the area see J.B. Peires, 'Facta non verba: Towards a History of Black Rugby in the Eastern Cape' (unpublished paper, University of the Witwatersrand History Workshop, 1981).
28. 'Umanyano nge mfundo', *Isigidimi Sama Xhosa*, 1 Feb. 1884.
29. On this see B. Willan, 'An African in Kimberley: Sol T. Plaatje, 1894–8' in S. Marks and R. Rathbone (eds.), *Industrialisation and Social Change in South Africa, African Class Formation, Culture and Consciousness, 1870–1930* (London, 1982), pp. 250–2.
30. Editorial notes, *Imvo Zabantsundu*, 28 Feb. 1889.
31. 'Natives and cricket', *Imvo Zabantsundu*, 23 March 1885.
32. Editorial notes, *Imvo Zabantsundu*, 9 March 1885.
33. Editorial notes, *Imvo Zabantsundu*, 3 Nov. 1884.
34. Marks and Rathbone, *op. cit.*, pp. 248–9.
35. 'Ibala labadlali', *Imvo Zabantsundu*, 21 Jan. 1897.
36. 'Obscene practices', *Imvo Zabantsundu*, 30 July 1891.
37. 'A sound mind in a sound body', APO, 15 Jan. 1910.
38. D. Kelly, 'A Historical Survey of East African Cricket', *The Cricket Quarterly* II (1964), 4–10, and J. Ferguson, 'Cricket in Nigeria', *The Cricket Quarterly* IV (1966), 225, 233–238.
39. J. Morris, *The Spectacle of Empire: Style, Effect and the Pax Britannica* (London, 1982), Chap. 8.
40. Quoted in R. Ross (ed.), *Racism and Colonialism* (Leiden, 1982), p. 121.
41. Quoted in J.P. McKay, B.D. Hill and J. Buckler (eds.), *A History of Western Society, II: From absolutism to the Present* (Boston, 1983), p. 930.
42. Quoted in Archer and Bouillon, *op. cit.*, p. 18.
43. A. Odendaal, *Cricket in Isolation : The Politics of Race and Cricket in South Africa* (Cape Town, 1977), p. 325.
44. Editorial notes, *Imvo Zabantsundu*, 30 March 1885.
45. For example, 'An oppressive municipality', *Imvo Zabantsundu*, 25 Nov. 1897; Cape Archives 3/KWT, 4/1/95: Memorandum from town clerk to borough ranger and forester, King Williams Town, 6.9.1911; Cape Archives, NA 636, file no 2207, statement by J. Jones re. cricketers ejected from native location, Uitvlugt, 9.1.1904.
46. G.19-1909 Blue Book on Native Affairs 1908, p. 25.
47. For an historical overview see A. Odendaal, 'History is on our side: Historical perspectives on constitutional change and African responses in South Africa' (paper no 46, conference on economic development and racial domination, University of the Western Cape, October 1984).
48. On this see S. Marks and A. Atmore (eds.), *Economy and Society in Pre-industrial South Africa* (London, 1980), pp. 27, 246–74 and Marks and Rathbone (eds.), op. cit., pp. 27–8.
49. For a detailed discussion of the social life and activities of the black élite in Kimberley see Marks and Rathbone, *op. cit.*, Ch. 9 and B. Willan, *Sol Plaatje: A Biography* (London and Johannesburg, 1984), Ch. 2.
50. 'Ibala labadlali', *Imvo Zabantsundu*, 18 Jan. 1888.
51. 'Ibala labadlali', *Imvo Zabantsundu*, 26 Aug. 1897 and 9 Dec. 1897; 'A Rhodes cup', *Imvo Zabantsundu*, 29 Sept. 1897; 'S.A. coloured rugby football', *Imvo Zabantsundu*, 2 Dec. 1897.
52. 'A Barnato trophy' and 'The Barnato memorial trophy', *Imvo Zabantsundu*, 2 Dec. 1897.
53. '"Imvo" special wires' and 'Ukuzigcobisa', *Imvo Zabantsundu*, 12 Jan. 1899.
54. Odendaal, *Cricket in Isolation*, p. 329.

55. C. Van Onselen, *Chibaro: African Mine Labour in Southern Rhodesia, 1900–1933* (Johannesburg, 1976), pp. 190–1.
56. See, for example, 'Kwa Mzilikazi', *Imvo Zabantsundu*, 4 Sept. 1899.
57. ' "Imvo" special wires', *Imvo Zabantsundu*, 12 Jan. 1899.
58. Hattersley, *Social History*, p. 115.
59. Odendaal, *Cricket in Isolation*, p. 325.
60. P.F. Warner, *The M.C.C. in South Africa* (Cape Town, 1906), p. 2.
61. See, for example, P.F. Warner, *Cricket in Many Climes* (London, 1900), p. 192.
62. R.E. van der Ross, 'The political and social history of the Cape coloured people, 1880–1970', part 3 (unpublished manuscript), p. 626.
63. *Indian Opinion*, 6 March 1909, p. 104. These and similar laws remained in force when the Transvaal became a British colony after the Anglo-Boer War.
64. See Berliner Missionsberichte, 1894, p. 291. In 1893, Rev Köhler of the Berlin Missionary Society at Potchetstroom forbade his black congregation from participating in a cricket match at neighbouring Klerksdorp because he feared that they might get out of control.
65. A. Odendaal, *Cricket in Isolation*, pp. 307, 328–9; 'Ukuzigcobisa', *Imvo Zabantsundu*, 18 June 1898; 'Ibala labadlali', *Imvo Zabantsundi*, 6 Feb. 1890 and 7 Jan. 1897.
66. On the unique history of this area and the political activities and social composition of the population see C. Murray, 'Dispossession and relocation in the Thaba Nchu district: A proposal for a regional study in the eastern OFS' (University of Cape Town, Centre for African Studies, Africa seminar, 16 Sept. 1981) and A. Odendaal, *Black Protest Politics*, pp. 21, 108–9, 158, 265–6.
67. 'Colour and sports', *Tsala ea Batho*, 19 Aug. 1911.
68. On the social stratification within the Indian community at this time see M. Swan, *Gandhi: the South African Experience* (Johannesburg, 1985).
69. Van der Ross, op. cit., p. 624.

12

Social Darwinism, Private Schooling and Sport in Victorian and Edwardian Canada[1]

David W. Brown

The publication of Charles Darwin's *On the Origin of Species* in 1859 had, of course, a marked impact on the Victorians. His fresh appraisal of evolution by the process of natural selection appealed to and sometimes appalled late nineteenth-century intellectuals and social reformers. Beside establishing 'a new approach' to 'the conception of development', Darwin's theory also shattered traditional Victorian beliefs.[2] Rejection of the evolutionary hypothesis on moral and religious grounds paralleled acceptance of the doctrine at a working and rhetorical level. The popular dictum 'the survival of the fittest', albeit an over-simplified interpretation of Darwin's ideas, provided a convenient and comprehensible explanation to many of the struggle for existence which was apparent in many spheres of life. Throughout the English-speaking world, philosophers, politicians and militarists now considered wider social ideas and issues from a Darwinian standpoint. Notable among those who gave credibility to the new evolutionary theory was Herbert Spencer. His application of the biological scheme of evolution to the realms of social relations gained an eager following in Britain and then overseas. Americans in particular were greatly influenced by Spencerian sociological thought.[3] William Graham Sumner became a committed disciple as well as a 'vigorous and influential Social Darwinist'.[4] Across the American border in Britain's 'Dominion of the North', Canadians, although 'comparatively secluded . . . from the sway of the vast controversy',[5] also involved themselves in the Darwinian debate.

Victorian Canadians, as intellectual historian Carl Berger has noted, 'were not original in transposing the doctrines of biological evolution' to account for variations in the social structure of society.[6] They applied the 'vulgarized conception' of Darwin's hypothesis freely. In fact, such was the impact of Darwin's work that Goldwin Smith, an eminent historian, journalist and political critic, suggested that it, or rather the misinterpretation of it, was 'running mad'.[7] Some commentators discussed publicly the logic of the metaphoric extension of the theory to certain aspects of social phenomenon, including religion and morality.[8] Others employed Darwinian rhetoric to explain social and racial inequalities.[9] Indeed, many issues received Darwinian consideration. And education and sport were not exempt from this potent new doctrine which so significantly influenced middle- and upper-class attitudes to life and society in the Victorian and Edwardian period.

This idea has been developed considerably with reference to British upper-class education in Mangan's reappraisal of conditions at the prestigious public schools.[10] Orthodox views of these élite establishments in the post-Arnoldian era have favoured the notion that the system spawned a continuous line of products bearing the qualities of the 'Christian Gentleman'. The schools certainly promoted this idea. They were ably assisted in this regard after 1857 by the 'copper-bottomed mould' for the English public schoolboy, that 'robust' and healthy 'muscular Christian', Tom Brown.[11] However, Mangan has argued convincingly that muscular Christianity has provided a far too simple explanation for the underlying aims of a public school education. The concept did help to erase the rough view of the pre-Clarendon Commission schools and facilitated the growth of a powerful educational ideology, athleticism, to which school authorities consistently subscribed.[12] But, 'life at these schools frequently owed little to the Christian Values and can be better understood by reference to a simplistically decoded Darwinian interpretation of existence which harmonised with wider social values of the time'.[13] In their efforts to equip their wards with 'character', and in the methods of its inculcation, the public schools were indebted as much to the tenets of Social Darwinism as to the principles of muscular Christianity.

Victorians were obsessed with the concept of 'character'. Moral earnestness and compassionate gentility symbolized ideal qualities which it was deemed necessary for men to possess if they were to be perceived as 'men of character' and 'decent' members of society. These qualities were central to living a Christian life. As one historian on the Victorian 'frame of mind' has observed, 'to be an earnest Christian demanded a tremendous effort to shape the character in the image of Christ'.[14] Christ-like attributes such as a sense of duty, virtue, courage, self-discipline and purity, were, of course, encapsulated in the 'ultimate masculine quality' of manliness.[15] And a man's manliness was tested in everyday life and enhanced by involvement in approved physical activities. In the second half of the nineteenth century, Thomas Hughes' 'muscular Christianity' and Charles

Kingsley's 'manly Christianity' augmented Thomas Arnold's vision of manliness which endorsed the Christianizing of men's natures through moral endeavour. Certain sports, and notably games such as cricket and football, were viewed as 'manly', and considered worthy for the training of a man's body 'for the protection of the weak, and the advancement of all righteous causes'.[16] In consequence, Victorian manly sports were embraced with fervour in Britain and the Dominion where conscientious disseminators of British culture were keen to reproduce the best cultural traditions of the 'Mother Country'.[17]

It has been noted that in their efforts to broaden the base of Empire overseas, 'the British were heavily dependent on their own imported inventions for keeping up morale'.[18] Reproduction of economic, political, religious, educational and social institutions played a crucial role in the re-establishment, maintenance and propagation of a highly-prized culture. It is not surprising to find therefore, that English-speaking Protestant settlers in British North America, and Canada after Confederation in 1867, adopted the basic elements of British public-school education. Throughout the nineteenth and early twentieth centuries, schools such as Upper Canada College, Toronto, Ontario (1829), Trinity College School, Port Hope, Ontario (1865), St John's College School, Winnipeg, Manitoba (1866), Rothesay Collegiate School, Rothesay, New Brunswick (1877), Lakefield College School, Lakefield, Ontario (1879), Bishop Ridley College, St Catharines, Ontario (1889), St Andrew's College, Aurora, Ontario (1899), University School, Victoria, British Columbia (1906) and Lower Canada College, Montreal, Quebec (1909) promoted with conviction an education based on Christian morality and character formation in turning out future citizens of true 'Christian manhood'. The headmasters of these schools were the most ardent proselytes of the 'techniques of persuasion and instruction' utilized by their British counterparts. Organized sport, sustained by the ideological rationalization of athleticism, was used intentionally as a means by which to nurture character.

Athleticism in the public schools of Victorian and Edwardian Britain represented in part a 'fusion of ideals;' traditional Christian motives existed in conjunction with Darwinian effect.[19] Jean Barman also indicates the co-existence of these two seemingly contradictory doctrines in private education in the Dominion of Canada.[20] A description of living conditions at one of these Canadian élite schools of the same period reinforces Barman's claim. One of the renowned schools of Ontario, Trinity College School, was depicted to the general public in 1890 as follows: 'The school premises now consist of more than twenty acres of land, on which has been centred a handsome and large building, including a beautiful chapel, presenting a south-front of eighty feet, warmed throughout with steam and hot air and lighted with gas and electric light'.[21] This comforting picture, however, omitted certain realities. A former schoolboy of the era viewed the College and its buildings in a rather different light. As Mangan has suggested with reference to the British public schools, there was an image for public consumption and one for private practice: 'When I entered the School it was comparatively small and the conditions were very primitive. In fact, no present day mother who has any regard for her boy and no family physician would today approve of a boy attending an institution if conditions were as primitive and insanitary as were at the School fifty years ago'.[22]

This unhealthy state of affairs was accentuated by further conditions of hardship: Canada's severe northern climate and participation in organized and compulsory physical activity. The resulting product epitomized more than the school authorities' intended outcome of the Christian gentleman. He was frequently the 'fittest' possible Christian gentleman. Darwinism and Christianity forged a contradictory, yet effective, alliance, in both ideal and practice, within the confines of the upper-class schools of Victorian and Edwardian Canada.

Athleticism in the Canadian private schools has been defined as an ideology which at the practical level comprised considerable and often compulsory involvement in organized physical pursuits, especially the games of cricket, rugby, Canadian football and ice hockey, and the activities of snowshoeing, tobogganing, skating, scouting, camping and canoeing. By taking part in such exercise, it was believed that an individual developed physical and moral qualities of benefit in later life. A sense of duty to self and group, which could include a team, the school and ultimately the nation, the attributes of honesty, truth and fair play, and the ability to co-operate, command and obey, were effectively transmitted through participation.[23] A number of nineteenth-century social and intellectual forces underpinned this educational ideology, providing its adherents with a coherent set of arguments upon which organized and compulsory physical activity could be justified. Christian evangelists gave athleticism a muscular Christian dimension. There were many in the Dominion who 'admired the doctrine of muscular Christianity' set forth in *Tom Brown at Rugby and Oxford*.[24] Educationists in general admitted openly the debt that 'physical individualism' owed to Hughes' ideal and to Kingsley's 'healthy animalism'.[25] And private-school adherents, in particular, clung tenaciously to the Christianizing tenets of athleticism and also espoused its nationalistic and imperialistic associations. In this latter regard, sport played a rather conflicting role: 'certain forms of physical activity were illustrative of Canadian character, and yet, in terms of the benefits derived from participation, games, and especially the maintenance of cricket, they were symbolic of the British sense of identity and the link with Empire'.[26]

Christian, nationalistic and imperialistic ideas were prominent features in the pedagogical rhetoric at Canada's élite schools. As was the habit in the public schools of Britain, headmasters, whether of British or Canadian origin, expressed themselves frequently in moralistic Christian tones. However, Darwinian undertones were often hidden beneath such orthodox terms of reference. The existence, therefore, of Darwinian activity within these institutions is paradoxical, for its underlying principles were not compatible with those of Christianity. Nevertheless, Darwinism existed and surfaced in the realities of schoolboy life.

Institutional Darwinism, associated with British public schooling, has been defined by Mangan at one level as 'the cultivation of physical and psychological stamina at school in preparation for the rigour of imperial duty'.[27] In Canada's élite schools, a similar sense of duty to the Empire was evident. And for the purpose of this discussion, the nature of institutional Darwinism in the Dominion will also be regarded as the intentional development of physical, moral and mental strength at school in preparation for the adversities of later life. This aim was an integral component of the character-building process to which private school educators subscribed. Conformity to the ideal was tediously consistent. Robert Machray at St John's College School (1866–1903), Charles Sanderson Fosbery at Lower Canada College (1909–35), Charles Bethune at Trinity College School (1870–99), John Ormsby Miller at Ridley (1889–1921), George Exton Lloyd at Rothesay Collegiate School (1891–96) and George Parkin at Upper Canada College (1895–1902) can be numbered among the many headmasters who joined forces to turn out 'manly, Christian and capable men', fit to 'successfully fight life's battle in the world'. To this end, they relied heavily on proven programmes established in Britain's public schools, notably the use of 'the chapel, the schoolroom, and the playground' to bring about the harmonious development of the 'tripartite nature of the boy, spirit, mind and body'.[28] The methods and success of Canadian private school education were summarized succinctly by H. Symonds of Trinity College School in the early 1900s: '*Anglo-Saxon superiority* is not the product of book-learning alone but of many factors, qualities of heart and head and *stout spirit*, which have been cultivated in body and school life. . . .'.[29] But 'Anglo-Saxon superiority' and 'stout spirit' owed little, in reality, to the religious facet of school life and school ideals. What headmasters chose to see as a Christian environment was in fact one of hunger, brutality and austerity.

Physical hardship was the lot of Canada's 'Tom Browns' in all aspects of schoolboy life. Living conditions were uncomfortable and severe disciplinary measures were administered frequently and rigidly. All were designed to build character. Thomas Aldwell recalled days at Trinity College School during the 1870s: 'The routine was necessarily strict. . . . Food, while adequate, was not wasted, and our appetites were those of growing boys. Often I spun knives with other boys for toast at breakfast. If I won, I was not hungry that morning; if I lost, I was a little hungrier than usual. It was worth taking a chance . . . *it was survival of the fittest* . . . it was a real boys' training'.[30] Other memoirs also reveal vivid pictures of school conditions and indicate that the world of the schoolboy held its 'fears' and 'dreads' as well as its 'fond memories'.

The cruel and sadistic treatment of boys by boys at Britain's public schools has been well-documented by Mangan.[31] While from available accounts, occurrences were less severe and less numerous than those in Britain, incidents of cruelty were to be found in Canadian schools. In the early 1900s, Raymond Massey, the noted stage and screen actor, attended first Upper Canada College Preparatory School and later St Andrew's College. His opinion of self-regulation by the boys at Upper Canada is succinct. It was, he stated, 'as bad as lynch law'. At St Andrew's, things progressed to 'mob sadism at its worst', and the description of the punishment labelled the 'sweats', during which the 'guilty' party was subjected to periods of suffocation under piles of mattresses, proves his point.[32] Peer group discipline was organized and, to a degree, premeditated. It was also institutionalized. For example, Trinity College School prefects were considered an extension of the headmaster's authority; they were permitted to oversee fighting between boys even though 'the rule was that there should be no fighting'.[33] At other schools across the country, prefects and senior boys were given varying degrees of discretionary power. The attitude of the masters was not uncommonly any less compassionate. One former student at St John's College School in Winnipeg in the 1890s recalled:

The discipline was rigorous. No time was wasted on such notions as self-expression and the like. If you failed a Latin exercise you were flogged for it, and I have a vivid recollection that after a flogging one did not fail any more. There may have been some boys with certain intellectual weaknesses who received special consideration, but the general rule was that for any inattention or misbehaviour, a good licking was certain.[34]

Charles Camsell, another Johnian, was in complete agreement. 'Life at St. John's', he remembered of his days in the 1880s, 'was pretty rugged', and, he added, *'the boys had to be taught to survive'*.[35] John Morgan Gray, a leading figure in the Canadian publishing world, spent a number of years in his early youth at Lakefield College School. He remembered in great detail the barrack-room hardiness of 'the Grove'. The 'smell of urine from the permanently impregnated floors' due to spillings from bed pots, was ever present. In winter, the situation worsened as 'the jerries were usually frozen solid and the floor was like a skating rink'. Windows were open to the elements until the rising bell sounded in the morning when compulsory cold baths and lakeside immersions rudely awakened the school's youth. Cleanliness may have been next to godliness, but this popular maxim could not cloud, even in the eyes of the boys, the intent: '. . . the object was hardiness'.[36] Lakefield, said Gray, was 'a hard school . . . but not cruel'.[37]

Ostensibly, these private schools were self-professed training grounds for the Dominion's future citizens – citizens of true Christian character. However, as we have seen, life was harsh and appeared on occasion to be un-Christian. This apparent contradiction can be explained. Headmasters were able to reconcile the value and content of their type of education with Christianity. Because of personal experience of private schooling, or knowledge of proven and celebrated methods, they remained conservative in their approach. They chose not to see the stoic nature of their system. And perhaps, like George Parkin of Upper Canada college, they had little trouble 'accommodating Darwinian science' with their Christian outlooks.[38] To men like Parkin, the process of evolution explained the history of life, not its origins; it was 'simply God's method . . . God working through nature'.[39] Struggles at school were merely aiding the realization of the school's Christian mission. From available recollections, the boys were aware of the authorities' stated objectives. But ideal and reality, in their eyes, did not necessarily correspond. For some, survival within the system required, primarily, a certain innate physical hardiness. For others, acquiring the physical, and psychological attributes was even more necessary to overcome the actual living conditions. In many ways, the way of life for the boys was 'Sparto-Christian' in nature.

Sparto-Christianity was central to the concept of manliness which evolved in nineteenth-century Britain, with the famous Scottish headmaster, H.H. Almond of Loretto School, leading the way in its advocacy. In Arnoldian terms, manliness fostered self-reliance and unselfishness through Christian and moral service. The association between muscularity and morality, between games and Christian codes of behaviour, and between trials of physical endurance and 'Sparto-Christianity' superseded this 'Arnoldian' outlook. An initial source of inspiration for this ideal stemmed from the social criticism of Herbert Spencer. Almond openly admired Spencer's 'laws of life' outlined forcefully in *Education: Intellectual, Moral and Physical*, published in 1861. Spencerian functionalism 'furnished Almond with a powerful rationale for his own physiological Darwinism',[40] which gained wider appeal in the British public-school system through the concept of 'sturdy sporting manliness'.[41] Across Canada, Spencer's views were recognized and publicized in popular and educational circles. In 1875, C. Clarkson echoed Spencer's imperatives in *The New Dominion Monthly*:

> The cry is for men – not mere helpless calculating machines, or animated classical dictionaries . . . it is of immensely more importance for a man to know . . . the chemistry and physics of everyday life, the physiology of his own body, and the laws of health and disease . . . than to be familiar with Greek tragedy or with quaternions, and know almost nothing about the simplest plant or animal, or even the very sunbeam that enters the eye.[42]

Canadians, in Spencerian fashion, were also advised of the benefits of clothing and exercise, with Medical Practioner, Dr J. Milner Fothergill, arguing the physiological advantages of the latter in 'Health in Childhood'. The 'perfection of health' by activating the muscles was responsible for 'the growth of healthy bones, ligaments and nerves' as well as for the development of the chest and the enlargement of the lungs.[43] Similar expressions of the 'general laws of health' were proclaimed by Charles Fosbery at Lower Canada College in Montreal. Fosbery took Spencer at his word, quoting his views regarding hygiene and health, and, like Almond, he extolled the virtues of all-weather exercise, cleanliness, comfortable dress and fresh air.[44] Fosbery was one of a number of Canadians who believed unequivocally in the character-building qualities of the country's natural elements.[45] In Canada, or this 'Brighter Britain', as George Dickson, headmaster of Upper Canada College in the early 1890s, designated it, the climatic conditions amplified the physical and psychological benefits to be gained from a physically demanding lifestyle. Over a decade earlier, another George, George Beers, was even more forceful in his advocacy of Canada's geophysical heritage. Beers, often called the 'Father of Lacrosse' for his zealous promotion of this indigenous game, linked climate and sport skilfully with the common denominator of character-building:

> The blood-born and bone-bred love of open air sports is the most marked physical characteristic of the Anglo-Saxon heritage in Canada. Nowhere under the sun is the climate more favorable for the hardy exercises in which the English-speaking people delight to indulge . . . I think the Canadians well typify the hardiness of the northern races; and nothing has perhaps helped more to form the physique of the people than the instinctive love for outdoor life and exercise in the bracing spring, winter and fall of the year.[46]

Other writers, intellectuals and educationalist of the era re-emphasized Beers' observation.[47]

Alick Mackenzie, headmaster of Lakefield College School from 1894 to 1938, at all times ensured that his boys experienced the natural surrounds of his rural school. Camping, snowshoeing, skating, tobogganing, swimming and canoeing each had their virtues and the special appeal that they could be practised within a stone's throw from the school building. In 1910, a contrived letter from an Englishman, John Bull, appeared in Lakefield's journal, *The Grove Chronicle*. After describing life on the lake and in the woods, the writer concluded: 'I couldn't help wondering if all Canadian boys had such a good time and why we English pitied those who lived in such a severe climate. I saw nothing but rosy cheeks and enjoyment in the frosty air'.[48] The sight of Lakefield's rosy-cheeked youngsters would have delighted a more renowned advocate of the concept which Carl Berger has termed the 'northern character' ideal. George Parkin, biographer of Edward Thring and later Trustee of the Rhodes Scholarship, was convinced of Canada's superior geophysical attributes. He was convinced that the climate produced a stronger and healthier people. Darwinian phrases were evident in his ejaculations on the topic. The harsh climate squeezed out the unfit and created 'a hardy race of the north'.[49] Another headmaster of the area, John Ormsby Miller of Ridley, was of a similar view, and noted that Canada was 'rich in soil, in natural wealth, in fertility of its soil, and all that goes to nourish a vigorous and progressive race'.[50] And Charles Gordon, a former master of Upper Canada College and better known as Canada's popular novelist, Ralph Connor, implied that 'nature's forces' could even re-shape the gentlemen emigrants who invaded the Dominion's shores from Britain's public schools throughout the nineteenth century.[51] In *Corporal Cameron*, Allan Cameron a former international rugby player for Scotland, is improved, both physically and mentally by the harsh Canadian environment: 'There was no "porsch" or sign of one on Cameron's lean and muscular frame. The daily battle with winter's frost and blizzards, the hard food had done their work on him. Strong, firm-knit, clean and sound, hard and fit, he had lived through his first Canadian winter. . . . Never in the days of his finest training was he as fit to get the best out of himself as now'.[52]

An atmosphere of robust living prevailed in the struggle to overcome the forces of nature as well as the austerity of school conditions. Sustained exercise augmented this rigorous atmosphere. Thomas Aldwell on athletics at Trinity College School: 'Games were important: football, foot-races, and cricket. But again there was no coddling. *A boy had to fight for survival*, had to learn to run and dodge and put everything he had into the contest or he was marked a failure'.[53]

And at St John's, Winnipeg, before the First World War, one Johnian recalled:

> It was a tough school morally and physically. Here I experienced the anxiety and satisfaction of taking part in the organized games which our elders then considered essential in the preparing the young for the exertions and temptations of adult life. Fair or foul, hot or cold, a goodly portion of our days was devoted to rugby and hockey under the stern supervision of our seniors. There were various informal but no less violent varities of physical competition to occupy our few unscheduled hours. As the terms went by, my endurance increased with my stature, and my fear of exhibiting embarrassing incompetence diminished with experience.[54]

Subscription to organized games in the Victorian and Edwardian private schools of Canada was consistent and widespread by the 1890s.[55] Games were lauded for their contribution to character-building. They helped to temper an individual's physical, mental and moral faculties. They were rationalized as being an integral means of nurturing the desired attributes of courage, endurance, obedience, modesty and self-discipline in the individual, and ultimately, the nation. Success in games was equated with ultimate success in life. 'Every encouragement' was given to 'genuine sport of whatever kind' in the belief that they were 'useful in the training' of the boys for the struggles in later life.[56] This notion was not just evident in the domain of the private school. *The Dominion Illustrated Monthly*, in 1892,

eulogized rugby football for its cultivation of 'pluck and determination', and noted that 'grit cultivated by the hard knocks of the football
field will stand the men in good stead in the contests of business or
professional life'.[57] The article also illustrated quite clearly the
association between this 'manly sport' and 'Saxon nature'. Institutional Darwinism and imperial Darwinism represented a healthy
alliance in the vocabulary of journalists and educators alike.

Private school pedagogues were united in their advocacy of the
athletic ideal and its associated values, but their emphases were
frequently disparate. At Ridley College, John Ormsby Miller was as
committed to publicizing the benefits of physical exercise as any of his
British or Canadian counterparts. The enduring basis for Miller's
educational ideals was a devout Christianity. His aim at Ridley was to
develop young citizens of virtuous standing, shaped in the image of
Jesus Christ, who, in Miller's own words was the 'model of all virtues
. . . the example to the human race of all the traits of true manliness
which men admire'.[58] Within the parameters of this imperative, Miller
defended the need for bodily exercise. In *Short Studies in Ethics*, his
justification reflected the literature of Carlyle, Ruskin, Hughes and
Johnson; and it clearly demonstrated the influence of Charles
Darwin. In the allegorical tale of the two canoeists in the Mozambique
Channel, it is the stronger of the two who survives an accident, 'an
example', Miller stated, 'of the difference wrought in two men merely
by exercise, or the steadiness of training'.[59] It was the commitment to
daily physical 'grooming', he argued, which made the body strong and
which produced the immense energy which characterized the English
race and which prompted 'the growth of vast colonies' and 'the
maintenance of empire over less civilized peoples'.[60] Imperial
Darwinism flourished in Miller's panegyrics on Christian morality. In
fact, his beliefs represented a triple fusion of ideals: Christian
gentility, institutional Darwinism and imperial Darwinism.[61] This
ideological triad flourished with varying degrees of emphasis on each
of the doctrines in the rhetoric and literature of athleticism of
Canada's private schools. This was particularly true of the Great War
period when the 'Call to Imperial Duty' proliferated in school
magazines.

Application of Darwinian theory to imperial duty reflected the
belief in the 'God-granted right' of the Anglo-Saxon race to conquer
the 'inferior' Teutonic opponent. While this belligerent attitude concerning the superiority of the Anglo-Saxon people, and its association
with racism, imperialism and Darwinism, was popular during the First
World War years, it was also prevalent in the intellectual and
educational circles at the turn of the century.[62] In 1909, speaking

before the Faculty of Education at the University of Toronto, William Hamilton Merritt, an Old Upper Canada Collegian, President of the Canadian Military Institute (1905–14) and 'undoubtedly the most active spokesman for military preparedness' in the Dominion, explained the role of war in social evolutionary terms. 'Nature,' claimed Merritt, 'seems to regulate that the world shall not be overrun with one species.' He continued: 'The study of both biology and history show me that the weak will go to the wall. It always has been and always will be the survival of the fittest'.[63] Some ten years earlier, in the 1890s, George Parkin had employed Darwinian terminology to explain the racial and political views on the state of society. Robert Page has written that Parkin 'viewed the world situation as a struggle between nations and races for survival and it would be survival of the fittest'. In Parkin's own mind, 'the fittest were the Anglo-Saxons'.[64]

Roughly coinciding with Parkin's pronouncement, one of the country's national journals, *The Canadian Magazine*, discussed the merits and strengths of the Anglo-Saxon race in articles such as 'Anglo-Saxon Superiority' and 'Saxon or Slav: England or Russia?'[65] The 'Call of the Empire' after Britain's entry into hostilities in August 1914 amplified this theme, and sport functioned conveniently as a potent medium through which similar messages could be delivered. One proselyte at Upper Canada College in 1915 considered athletics from the viewpoint that they 'are only to build us physically, and to get us fit to play our part in the struggle for the Empire's existence . . . as only a means of enabling us to avenge our glorious dead'.[66] This conviction, based on physical supremacy, was reiterated some two years later at another Ontario institution. James Hughes, perhaps the most prominent figure in Canadian public education after Egerton Ryerson, preached to Trinity College schoolboys that 'games . . . build character' and that they were responsible for 'that never-say-die spirit – that spirit that saved the day for the allies at Ypres'.[67] Athleticism, from available accounts, was the esteemed consort of militarism. Games maximized the physical development of the individual, and ultimately, the nation and Empire, for the struggle of war.

While it is beyond the scope of this investigation to deal more comprehensively with the application of Darwinism to the concepts of racism, militarism and imperialism in the Great War years, the identification of its manifestation is worthy of note as it is indicative of the complex system that 'co-existed, openly competed and overlapped' in this educational setting.[68] As in Britain, the association between sport and war, and between the playing field and the battle-ground, was a salient feature in the literature of the Dominion's private schools. Schoolboy writers of poetry and prose portrayed physical and moral strength as superior Anglo-Saxon traits. They linked conflict in war with games, and implicated heroic battlefield deeds with imperial duty as many of their former colleagues gave 'their heart and soul for Anglo-Saxondom':[69]

> We have given our best, they have fought and bled:
> For the life of the Empire their lives were shed:
> To our lasting pride.
> They died.[70]

In his pioneering study on sport in Ontario private schools, Geoffrey Watson stated that the cultural trends of society at large exerted only a limited influence within the schools.[71] This past claim needs to be revised. Social and intellectual issues of the time found their way deep into the hallowed halls of these institutions. With general reference to nineteenth-century Ontario, Alan Metcalfe has demonstrated how divergent ideas permeated period sport and physical education. However, his analysis in one instance requires revision. Metcalfe concluded that the system of physical training in the schools was grounded in Darwinism and that organized games within the élite schools were indebted to the tenets of muscular Christianity.[72] This explanation was not wholly wrong, but the notion that muscular Christianity was the predominant influence in the private schools only comprises a partial analysis of the role of sport in this educational setting. Athleticism was an amalgam of ideals: Christianity, Darwinism, nationalism and imperialism. Darwinian sentiment and practice made a nonsense of simplistic Christian claims not only for games, but also for private school life. John Morgan Gray, the Lakefield product, remembered that during history class when the boys were introduced to the Greeks and encountered the Spartans, they 'recognized' themselves.[73] The Canadian private schoolboy was a paragon of 'virtue and manhood', but he was more than the embodiment of Christian manliness. He was a 'Spartan of the West'.[74]

NOTES

1. The author would like to thank the editors of *The Canadian Journal of History of Sport* for allowing parts of the article 'Sport, Darwinism and Canadian Private Schooling to 1918', in XVI, 1 (May 1985), 27–37, to be reproduced here.
2. See R. Hofstadter, *Social Darwinism in American Thought* (Boston, 1955).
3. Ibid., pp. 31–50.
4. Ibid., pp. 51–66.
5. Goldwin Smith, 'The Immortality of the Soul', *The Canadian Monthly and National Review*, 9, 5 (May 1876), 408–16.

 6. Carl Berger, *The Sense of Power: Studies in the Ideas of Canadian Imperialism, 1867–1914* (Toronto, 1970), p. 245.
 7. Elizabeth Wallace, *Goldwin Smith: Victorian Liberal* (Toronto, 1957), p. 194.
 8. See for example, John Watson, 'Science and Religion', *The Canadian Monthly and National Review*, 9, 5 (May 1876), 384–97, and 'Darwinism and Morality', 10, 4 (October 1876), 319–26; and J.A. Allen, 'The Evolution of Morality', *CMNR*, 11, 5 (May 1877), 490–501.
 9. See R.J.D. Page, 'Canada and the Imperial Idea in the Boer War Years', *Journal of Canadian Studies* V, 1 (1970), 33–49.
10. See J.A. Mangan, 'Social Darwinism, Sport and English Upper-Class Education', *Stadion* VI (1982), 92–115.
11. Ibid., 93–4.
12. Ibid., 96–8.
13. Ibid., 110.
14. Walter E. Houghton, *The Victorian Frame of Mind, 1830–1930* (London, 1957), p. 231.
15. Morris Mott, 'The British Protestant Pioneers and the Establishment of Manly Sports in Manitoba, 1870–1886', *Journal of Sport History*, 7, 3 (1980), 27.
16. Thomas Hughes, *Tom Brown at Oxford* (London, 1880), p. 99.
17. Mott, op. cit., 25–36; and Gerald Redmond, 'Diffusion in the Dominion: "Muscular Christianity" in Canada to 1914', paper presented at the Annual Conference of the *History of Education Society*, Loughborough, England, (1982).
18. R. Hyam, *Britain's Imperial Century, 1815–1914: A Study of Empire and Expansion* (London, 1976), p. 150; L.B. Wright, *Culture on the Moving Frontier* (New York, 1961) pp. 20–21; and D. Walker Howe, *Victorian America* (Pennsylvania, 1976), pp. 24–5. With reference to the transfer of British public school ideals to Canada, see David W. Brown, 'Athleticism in Selected Canadian Private Schools for Boys to 1918' (unpublished PhD dissertation, University of Alberta, 1984); Jean Barman, *Growing Up British in British Columbia: Boys in Private Schools* (Vancouver, 1984); and Patrick Dunae, *Gentlemen Emigrants: From the British Public School to the Canadian Frontier* (Vancouver, 1981).
19. J.A. Mangan, *Athleticism in the Victorian and Edwardian Public School: The Emergence and Consolidation of an Educational Ideal* (Cambridge, 1981), p. 136.
20. Jean Barman, 'Growing Up British in British Columbia: The Vernon Preparatory School, 1914–1946', in J.D. Wilson and D.C. Jones, *Schooling and Society in Twentieth-Century British Columbia* (Calgary, 1980), p. 126.
21. *The Dominion Illustrated, A Canadian Pictorial Weekly*, V, 90 (1890), 71.
22. Mangan, 'Social Darwinism', 96; and 'Recollections of R.C.H. Cassels ('89–'93)', *The Trinity College School Record*, 43, 4 (1940), 8–10.
23. Brown, op. cit., p. 201. This definition is adapted from Mangan's original study in 'Athleticism: A Case Study of the Evolution of an Educational Ideology', in B. Simon and I. Bradley, *The Victorian Public School: Studies in the Development of an Educational Institution* (Dublin, 1975), pp. 174–87.
24. *The Dominion Illustrated: A Canadian Pictorial Weekly*, VII, 168 (1891), 287. For a discussion of the promoters of the ideal in Canada, see Redmond, op. cit.
25. George M. Grant, 'Education and Co-Education', *Rose-Belford's Canadian Monthly and National Review* III (1879), 509–18; and Frances Cobble Power, 'The Little Health of Ladies', *The New Dominion Monthly*, April (1878), 443–58. It has been argued that muscular Christianity existed in the pre-literature of Hughes and Kingsley. See Gerald Redmond, 'The First Tom Brown's Schooldays: Origins and Evolution of "Muscular Christianity" in Children's Literature', *Quest*, 30 (1978), 4–28. For a fascinating analysis of the medieval concept of chivalry on this ideal and on the Victorian frame of mind and culture, read Mark Girouard, *The Return to Camelot; Chivalry and the English Gentleman* (New Haven, 1981).
26. Brown, op. cit., pp. 187–8.
27. Mangan, *Athleticism*, p. 136.
28. H. Symonds, 'The Boarding School System in Canada', *The Educational Monthly of Canada*, Jan. (1903), 21–6.

29. Ibid., emphasis mine.
30. Thomas Aldwell, *Conquering the Last Frontier* (Seattle, 1950), pp. 5–6, emphasis mine.
31. See Mangan, 'Social Darwinism'.
32. Raymond Massey, *When I Was Young* (Toronto, 1976), pp. 77–8.
33. An Old Boy's Reminiscences', *T.C.S. Record*, VIII, 2 (1905), 12–14.
34. W.A. Griesbach, *I Remember* (Toronto, 1946), p. 157.
35. C. Camsell, *Son of the North* (Toronto, 1954), p. 26, emphasis mine.
36. John Morgan Gray, *Fun Tomorrow: Learning to be a Publisher and Much Else* (Toronto, 1978), pp. 25–6.
37. Ibid.
38. Berger, op. cit., p. 233.
39. Ibid., p. 234.
40. Mangan, *Athleticism*, pp. 50–53.
41. Norman Vance, 'The Ideal of Manliness', in Simon and Bradley, op. cit., p. 128.
42. C. Clarkson, 'A Liberal Education', *The New Dominion Monthly (NDM)*, Oct. (1875), 262–3.
43. J. Milner Fothergill, 'Health in Childhood', *NDM*, Nov. (1875), 442–6; and 'Health and Dress', *NDM*, Oct. (1875), 274–6, which discusses the state and style of women's dress and their effects on health.
44. Charles S. Fosbery, 'Health', *The Eagle*, 1, 6 (1907), 56; and Brown, op. cit., pp. 71–4.
45. The notion of 'Northern Character' in Canadian intellectual thought is outlined excellently in Berger, op. cit., pp. 128–52.
46. W. George Beers, 'Canadian Sports', *The Century Magazine*, 14 (1877), 506–27.
47. See Berger, op. cit.
48. 'A Letter', *The Grove Chronicle*, V, 1 (1910), 8.
49. G.R. Parkin, 'The Educational Problems and Responsibilities of the Empire,' in J. Castell Hopkins, *Empire Club Speeches* (Toronto, 1912), pp. 70–80; and *The Great Dominion: Studies in Canada* (London, 1895), pp. 213–15.
50. John Ormsby Miller, *The Young Canadian Citizen: Studies in Ethics, Civics, and Economics* (Toronto, 1919), pp. 11–27.
51. See Dunae, op. cit., passim.
53. Ralph Connor, *Corporal Cameron* (New York, 1912), p. 309.
53. Aldwell, op. cit., emphasis mine.
54. Arnold Heeney, *The Things That Are Caesar's: Memoirs of a Canadian Public Servant* (Toronto, 1982), p. 18.
55. See Brown, op. cit., p. 37. It is incorrect to assume that a school founded before 1890 instituted immediately a system of organized games. Adoption could be either immediate or delayed.
56. E.M. Watson, 'Cricket at Trinity College School', *Athletic Life*, Jan. (1895) 51–2.
57. R. Tait McKenzie, 'Rugby Football in Canada', *The Dominion Illustrated Monthly*, 1, 1 (1892), 11–19.
58. John Ormsby Miller, 'Christian Education', An Address Delivered at Prize Day, 1890, Reprinted in *Bishop Ridley College Prospectus*, 1890, pp. 13–17; and *The Young Canadian Citizen*, p. 76.
59. John Ormsby Miller, *Short Studies in Ethics: An Elementary Text-Book for Schools* (Toronto, 1895), pp. 72–7.
60. Ibid., and *The Young Canadian Citizen*, pp. 5–6, 19–29, and 34–8.
61. See Mangan, *Athleticism*, p. 136. He elaborates this point. Imperial Darwinism is viewed as 'the God-granted right of the white man to rule, civilize and baptise the inferior coloured races'.
62. See Hofstadter, op. cit., pp. 170–200.
63. Berger, op. cit., pp. 59 and 246.
64. Page, op. cit.
65. D. Mills, 'Saxon or Slav: England or Russia?' *The Canadian Magazine* IV, 6 (1895), 518–539; and R.W. Shannon, 'Anglo-Saxon Superiority', *The Canadian Magazine* X, 4 (1898), 289–94.
66. U.C.C. and the War', *The College Times*, Summer (1915), 36.

67. *T.C.S. Record*, XIX, 3 (1917), 42–43.
68. Mangan, 'Athleticism', p. 166.
69. H.R. Rathbone, 'The Tie That Binds', in *Empire Club of Canada* (Toronto, 1919), p. 272.
70. 'Pro Patria', *The College Times*, Summer (1915), 3.
71. Geoffrey G. Watson, 'Sport and Games in Ontario Private Schools, 1830–1930', (unpublished MA thesis, University of Alberta, Edmonton, 1970), p. 40.
72. A. Metcalfe, 'Some Background Influences on Nineteenth-Century Sport and Physical Education', *The Canadian Journal of History of Sport and Physical Education*, V, 1 (May 1974), 62–73.
73. Gray, op. cit., p. 35.
74. This was a term used by Ernest Thompson Seton, the Toronto-based artist, naturalist, novelist and founder of the modern scouting movement in North America in the early decades of the twentieth century. The model for his scout as emulator of the backwoodsman, explorer and frontiersman was the native Indian, who in Seton's opinion followed the lifestyle of a 'Spartan of the West'. See E.T. Seton, *The Gospel of the Redman: An Indian Bible* (London, 1937). This was reproduced from *The Book of Woodcraft and Indian Lore* (New York, 1912).

13

Cricket and Colonialism in the English-Speaking Caribbean to 1914: Towards a Cultural Analysis

Brian Stoddart

In 1929 the American writer T.S. Stribling set an intriguing detective story in the world of Barbadian cricket.[1] The son of a prominent Bridgetown banker was found dead in the Wanderers Cricket Club pavilion shortly after helping that team win its toughest match of the season. Murder was suspected, a club professional the prime suspect. Bumbling investigations by an American university psychology professor, enjoying a Caribbean sabbatical, eventually revealed the villain as a respectable amateur member of the club. The story appears trivial. In its detail, however, it sets the foundations for what Michel Foucault might have termed an archaeology of Caribbean cricket.[2] Stribling located both the playing and the sociological conditions of cricket deep in Barbadian civic culture, seeing the game constituted by its social context rather than arising merely as an adjunct to it.

Stribling's insights were left at the superficial and unexplored level, but still provide invaluable clues for a social unravelling of Barbadian cricket history in particular and, by extension, of the English-speaking Caribbean in general. To his American eye cricket in Barbados was an integral part of local culture, prestige and power. That set him apart from earlier cricket-steeped writers like Pelham Warner and Lord Hawke; although both referred passingly to unique features of the Caribbean condition, neither provoke the analytical responses invited by Stribling.[3] (As C. Wright Mills pointed out, the unindoctrinated observer frequently seizes the possibilities of everyday material more readily than one socialized into the culture, be it national or sectional).[4] There is, then, a direct if curious link between Stribling and C.L.R. James who, in his magisterial 1963 work *Beyond A Boundary*, sought to reconcile a life dedicated to Marxism with a concurrent and contradictory obeisance to cricket and English literature, the antithesis of his radical ideology.[5] Since then a number of writers have essayed extensions to the James conclusions, but few if any have attempted an explanation of the construction process by which the conditions pointed to by Stribling and James came about.[6]

The purpose here, then, through a concentration on Barbados and with reference to the other major pre-1914 cricketing islands, is to isolate the underlying principles of the environmental culture of Caribbean cricket; the building of its social authority systems; the elaboration of its social classification process; the origins and growth of its peculiar professional tradition; and the development of its ideological acceptance and resistance practices. From this analysis, cricket emerges not simply as an agency for colonial recreation, but as an arena where players in the imperial and colonial 'game' struggled to decide the results of social proselytism and cultural power through sport.

By concentrating on Barbadian (or Bajan, as it is often referred to locally) cricket, Stribling confirmed the term 'West Indies' as a useful shorthand but a misguiding principle for social and historical investigation. Caribbean cricket is analysed most profitably through its specific components. There was certainly talk of West Indian cricket, political and other types of federation before 1914, but its non-arrival underlined the differences which distinguished the territories from their geographical and demographic conditions upwards.[7] British Guiana, at the Southern end of the Caribbean, occupied an area the size of Great Britain located on the South American mainland, much of it reachable only by water.[8] By 1891 East Indians (the regional term for indentured labourers from India and their descendants) outnumbered blacks, 122,000 to 113,000 with Portuguese 11,000 and other Europeans a slight 4,000. This varied population produced minerals, coffee, cotton and sugar. The island of Trinidad occupied about 1,900 square miles of very different country, had one of the most diverse populations on earth, as revealed beautifully in Edgar Mittelholzer's novel *A Morning At The Office*, and produced sugar, cocoa and oil. At the northern end of the region was Jamaica, over 4,000 square miles of mostly steep hill country and small coastal plains. By 1891 blacks there numbered 488,000, coloureds (the result of European and black intermixing, there being no significant East Indian population) 121,000 and whites 14,000. Its main crop was sugar with strong contributions from coffee and pimento. Apart from a heritage in slavery and an acquired British governance, these states varied considerably in economic and cultural composition which, along with the weather, produced distinctive cricket traditions.

From the beginning of inter-colonial competition in 1865 Jamaican cricket was sealed off from that in the other islands because of both distance and the irregularity and prohibitive cost of inter-island transport.[9] In the years to 1914 Jamaica saw more English touring teams than it did those from other colonies. Physical separation led to a deep seated psychological one so that there were frequent disputes over the organization of tournaments and, from 1900, the selection of combined teams to visit England.[10] In Trinidad weather conditions and attacks upon the soil by (ironically enough!) the mole cricket combined to encourage the use of coconut matting as a substitute for traditional turf pitches. Until 1935, when an Australian coach introduced jute matting, Trinidadian playing conditions were said to develop batsmen who were excellent cutters and hookers but not drivers, because the ball did not 'come on' to the bat.[11] In British Guiana, two wet and two dry seasons made co-ordinating a common playing time with the other islands difficult. By 1895 the playing rules for inter-colonial matches stipulated that they be completed between 1 August and 31 October, but experience showed even that to be impractical.[12] Among other things the intersection of these geographical, climatic and social statistics put home teams at a substantial advantage, and a win away was always a source of particular jubilation. The local specificity produced by these varying circumstances of environment is demonstrated especially well by the Barbadian case.

By the later nineteenth century the tiny, 166 square-mile island, popularly known as 'Little England', was crammed with over 180,000 inhabitants creating one of the highest population densities in the world.[13] Just over 15,000 were white, the rest largely black with a small but significant coloured community known, most commonly, as 'brownskins'. Unlike its cricketing partners, Barbados had known no other political allegiance but to Britain since the first settlers landed early in the seventeenth century, and one result was a representative government different from that in the other islands.[14] Representation itself was based directly on a unique political economy because Barbados, under imperial rule, had known nothing but the production of sugar; it had remained 'faithful to the cane'.[15] 'Crop', as sugar cane harvesting is still known, defined Bajan life as production requirements provided the impetus for slavery and, after abolition, the need to maintain cheap 'free' labour in the face of increasing industrial inefficiency, faltering prices and growing competition for a static market. Tough legislation such as the 1840 Contract Act, which created a located estate tenantry, determined that by the later nineteenth century 'working-class' had replaced 'slavery' in Bajan parlance.[16] But sugar went beyond even that. Its *idea*, its permeation of all aspects of island culture was as important as its economic *fact*. Every plane of Barbadian life was touched by sugar, cricket prominently included.[17]

Logically the game should have been played during the dry season but that was in 'Crop', so matches took place during the unpredictable wet months. Many 1877 games played by the newly formed Wanderers Club were disrupted by rain.[18] In 1895 the Barbados side routed the first touring English team for 48 runs in its first innings on a damp wicket, then won the match convincingly.[19] In a local cup match in 1896 the Pickwick Club made only 45 but still beat Leeward by 4 runs – over five inches of rain had fallen the previous day.[20] In 1905, another English touring team was dismissed for 95 in its second innings leaving Barbados just 20 to win, rain again the determinant.[21] Calls for reform of the season received little support because of the sugar culture, and from 1909 onwards local matches became even more unpredictable when their playing time was stretched from one Saturday to three.[22] Barbadians turned this uncertainty into a virtue, claiming it gave them experience of differing conditions unlike British Guiana players who, it was claimed, never played on wet pitches.[23] More important even than the impact of sugar upon playing conditions, however, was its determination of who was placed where in the social hierarchy of Bajan cricket.

Plantation owners, merchants, bankers, clerks, and civil servants came to competition matches categorized by their place in the hierarchy of sugar production, and cricket was a powerful instrument in the preservation and promotion of that hierarchy. Cricketers, for example, were conspicuously prominent in the Voters Lists until 1914 and beyond.[24] The significance is that for most of the period under review less than two per cent of the population qualified for a vote, the stipulations for which required either a substantial yearly income, the freehold of land, the payment of substantial parish taxes or the possession of a university degree. By definition these people were planters, professionals, merchants, bankers or senior clerks. Their influence was accentuated by constituency imbalances: in 1900, for example, the city of Bridgetown (which contained most of the emergent middle-class blacks) had 407 voters to return its two representatives while St. Lucy (a rural seat dominated by white planters) had just 47 voters to return its two. The aristocracy of cricket, then, was the aristocracy of sugar, but even that masks an important structural point.

By 1892, when the Barbados Challenge Cup competition was formalized, power within the sugar hierarchy was shifting rapidly from traditional plantation owners to a smaller group of merchants and associated service castes such as lawyers. As prices faltered and production costs climbed, owners unable to sustain operations were assisted by legislation such as the Agricultural Aids Act of 1889 under which merchants advanced loans against the imminent crop. By the early twentieth century sugar conditions were so poor that many planters were heavily indebted to merchants, and an increasing number were foreclosed by the lending institutions. Similarly, many estates changed hands in the Court of Chancery, basically an institution designed to pass properties from those who could not support them economically to those who could in order to keep the industry and, of course, the social economy going. Merchants were prominent purchasers, and towards 1914 the process was widespread.[25] In 1908, for example, a foreclosed planter who lost land held by his family for over 200 years attracted much sympathy, and the following year another traditional and influential planter severed his island connections.[26]

Cricket players and administrators (frequently one and the same) were prominent in this activity, men like[27] J.O. Wright (merchant, company director, planter, politician, Barbados Cricket Committee member and long-serving Wanderers President); D.C. da Costa (director of a family merchant firm, active cricket administrator and whose son, an excellent wicketkeeper, followed him into the business); A.S. Bryden (merchant, travelling agent and property dealer whose son, a Wanderers player, also took over the business); J. Gardiner Austin (founding member of Wanderers, merchant, politician and one of whose sons, H.G.B. Austin, became the doyen of Barbadian and Caribbean cricket as well as director of the reconstituted family company); and R.G. Challenor (a merchant whose son, George, became one of the greatest ever Caribbean batsmen as well as junior partner in the family firm which emerged at the end of the First World War as one of Barbados' largest trading conglomerates). These men were assisted by lawyers such as H.M. Cummins (long-serving secretary and player for the Spartan Club) whose practices thrived on the conveyancing of large property transfers. To speak of the 'plantocracy' dominating cricket, then, is imprecise. Although that group certainly provided numerous players and administrators, in the years down to 1914, the merchant-based emergent upper-middle class became increasingly influential. And for them, cricket became a social bulwark in the midst of alarming change which sprang from two inter-related conditions: a shaken faith in the British political as opposed to cultural connection, and a fear of the proletariat based on economic uncertainty.

In 1876 the so-called Federation Riots sprang from reactions to the

British government's apparent intention to alter the time-honoured Barbados constitution.[28] In the rioting, the élite divined that the 'working class', encouraged by the Governor, was beginning to demand greater rights.[29] Along with the urban growth of Bridgetown, this was interpreted as the old rural relationships breaking down along with the economic base of stability and good order. From the 1870s onwards there were endless complaints from 'respectable' society members, cricketers prominently among them, about the growing numbers of 'idlers and loafers' on the island, about the rise of youth gangs who terrorized and abused respectable citizens, about the insults and lewd suggestions offered to society women as they went through the streets, about the growing evils of grog shops and prostitution, and about the general increase in begging and lawlessness.[30]

One response to this unwelcome development was tough action by the lawyers and the judiciary.[31] In 1877 a field labourer received twelve months' imprisonment with hard labour for 'conspiring' to raise the wage of a cane-cutting contract. Setting fire to cane (among the gravest of anti-social acts) fetched up to 14 years with hard labour, and early in the twentieth century was thought in some quarters to warrant the recall of the cat-o'-nine-tails. In 1880 the Governor regretted a legal imprecision which seemed to prevent him from ordering female prisoners to have their hair cropped for breaches of discipline. In 1892 a woman received three months with hard labour for stealing two hens. By 1899 the Attorney-General thought 36 lashes not unreasonable for assault on girls under ten years. In 1909 an estate manager arbitrarily stopped wages for alleged worker indiscretions. When his employees disputed the decision with his bookkeeper, they were convicted on assault charges.

Among the judges and lawyers upholding this system were numerous cricket players and administrators: Sir Conrad Reeves (an extremely tough-minded Chief Justice and, from its foundation until his death, President of Spartan); G.A. Goodman (long-time Solicitor-General then Attorney-General, Barbados player and Pickwick President); Sir W.K. Chandler (Court of Appeal and Court of Chancery, President of Wanderers); H.S. Thorne (Police Magistrate and Barbados player); H.M. Seon (son of a Police Magistrate, court clerk, and Spartan player); Stephen Rudder (court clerk, Barbados and Spartan player); G.H. Corbin (solicitor and Spartan official) and E.C. Jackman (lawyer, planter, Barbados and Wanderers player). Cricket, then, had strong links with the legal arm of respectable society which was seeking to maintain civic order against what it considered massive social change.

.

A more subtle and, indeed, more influential defence against change than legal repression lay in cultural formation.[32] When Wanderers club was created in 1877, at least three of its officials had been key members of the Defence Association formed the previous year to oppose constitutional change: J. Gardiner Austin (sacked as a Governor's aide-de-camp for his actions); J.W. Carrington (a leading lawyer and politician), and T.B. Evelyn (attorney for the Bay estate from which the first Wanderers ground came).[33] D.C. da Costa and R. Challenor were other prominent Defence Association members also significant in early cricket circles. Cricket rapidly became a civilizing mechanism for this group, a source of both recreation and reaffirmation of social standards for its players and a display of civilized behaviour for spectators, especially those of the 'working class'. Cricket pitches were prominent in fashionable new housing developments like Belleville and Strathclyde which, from the 1880s, became monuments to the exclusivity of 'proper' society.[34] For the rising merchant group, in particular, this cultural style was an important means of maintaining social standards when older conventions were weakening. From these suburbs emanated hopes that 'propriety and decorum' might improve among the lower orders, preferably through 'an appreciation of the example which is set them by the higher classes'.[35] With the emerging popularity of cricket arose illusions of social unity which suggested the game transcended normal divisions of colour, class and status, even though it quite clearly preserved careful social distances within its organizational structures.

C.L.R. James is an invaluable analyst on this point as well as on the political role of Caribbean cricket more broadly, not least because he indicates both the relevance and power of Gramsci's theory of hegemony in such a context.[36] Gramsci argued, in general, that in many societies the ruled share with the rulers a belief in cultural values which over-ride economic and social inequalities. Consequently, opportunities for change are often lost because, for the ruled, to seize them would mean a rejection of those powerful shared cultural beliefs. James, who became a Marxist in spite of English literature and cricket, is the 'exception that proves the rule'. Before 1914 in Barbados, cricket was established as a major shared cultural value with the players being drawn from the élite, respectable section of the community, their exploits cheered on by the lower orders who themselves had few opportunities to play. These groups were part of a cultural authority system whose ideological model was drawn directly from the British heritage and its attendant ethical idealism through which cricket became as much moral metaphor as physical activity.

In the Stribling story Wanderers Cricket Club possessed a heritage dating to 1712 and a charter from the Prince of Wales. Historically the account is inaccurate but sentimentally Stribling was right, because the foundations of club cricket throughout the Caribbean were set upon a concern for British cultural continuity and social respectability. The social connotations of cricket reached far beyond the fences surrounding the Georgetown Cricket Club formed in British Guiana during 1852, the Kingston Cricket Club established in Jamaica during 1863 and Queen's Park Cricket Club in Trinidad, especially after its transfer to the Park from the Savannah early in the 1890s. In 1904 it was said that a 'respectable' visitor could *generally* be put up for temporary membership of Queen's Park, while a year later the Kingston Cricket Club was named 'one of the established institutions of the city'.[37] These clubs provided a cultural focus for the ruling élite which endeavoured to have its cricket and related values, themselves prized for their British role modelling, accepted by the rest of society.

As elsewhere in the British Empire, Caribbean cricket was based consciously on the reformed English game following the start of the county championship in 1873.[38] From at least the 1890s in Barbados, English and Australian scores and affairs were reported regularly and in detail by the newspapers. English manuals invariably were the authorities on technique, demeanour and conventions.[39] One concern was to demonstrate, through cricket, that English stock had not degenerated in the tropics.[40] Even though many cricketers came from Bajan families with island histories of 200 years or older, they still considered themselves British; cricket prowess was a yardstick of their success in remaining British, so victories against English teams produced great celebration.

The British attitudes aroused by cricket were far more important than its mere physical activities. There was simple loyalty to the Mother Country, an emphasized feature when any English team visited. Cricket symbolized wider unity. In 1895, for example, the affectionate farewell to the first England team was thought to reveal a bond 'other than a common enthusiasm for cricket'. G.A. Goodman noted simply that Barbadians were 'the sons of old England'.[41] In 1899 a spectator at Queen's Park Oval said that although Barbados had lost against Trinidad 'she would not give up because her sons were from old British stock'.[42] The extent and persistence of this sentiment was revealed in a 1923 speech by H.B.G. Austin to a Belfast cricket dinner during the West Indies tour of Great Britain. He hoped the tour would prove the West Indies worthy of the Empire; they did not want to lose their birthright and imperial privileges, he said.[43]

Then there was the symbolism of British statecraft through cricket. After Barbados beat British Guiana in 1891, one editorial comment offered as the social message of cricket which all citizens should imbibe: 'be loyal and obedient to constituted authority'.[44] A cricket team always 'elected' its captain, it was noted, then gave him undivided support even if he made mistakes. This replicated the British political system's democratic creation of authority and, like democracy, cricket created *esprit de corps* because team members entered a solemn contract to serve each other with maximum effort. This bred honourable behaviour because team members would not let each other down on or off the field. Consequently, 'respect for truth and right and fairplay is held in highest esteem'. So, when in 1909 against British Guiana a Barbados player ran himself out rather than the in-form George Challenor, the lawyer E.C. Jackman congratulated him publicly that 'In the true spirit of the game . . . he elected to sacrifice his own wicket'.[45] Conversely, lapses from the solemn, co-operative contract were most unwelcome. When the 1905 English tourists showed annoyance at losing their return match against Barbados, and when a team member criticized Barbadian crowds, their attitudes were received coolly.[46]

One heartland for the growth of this ideology lay in the prestigious schools on the island: Harrison College, The Lodge and, to a lesser extent, Combermere.[47] By 1900 over 24,000 island students were receiving education but these high schools were exclusive: Harrison had 154 students, Combermere 122 and The Lodge just 51.[48] Harrison was largely the preserve of the white plantation and mercantile élite, Combermere that of the growing black middle class although a few of these sent their children to Harrison, while Lodge, the only boarding school, tended to be less exclusive than Harrison. Harrison and Lodge both had teams in the Cup competition, although Lodge withdrew briefly early in the twentieth century – after a reformation early in the 1880s a string of headmasters and financial worries undercut it as a cricket power.[49] For most of the period its cricket fame rested upon the Goodman brothers until the appearance in the late 1890s of P.H. (Tim) Tarilton, one of the Caribbean's most technically correct and high-scoring batsmen of all time. Harrison College, meanwhile, grew considerably in strength.

It originated in 1733 as a school for poor and indigent boys, but was rejuvenated by an 1870 Act which established it as a grammar school for the benefit of the better classes.[50] Horace Deighton was appointed Headmaster in 1872 and, before his 1905 retirement, set the academic standard along with an English emphasis on the moral, social and cultural power of the playing field, especially the cricket ground. Dr Herbert Dalton then ran the school until 1922. He introduced the English school 'set' or 'house' system, and did much to raise school spirit on English public school lines. Both men were active on the Barbados Cricket Committee, their cricket influence in the school being boosted by their long service and by the presence of G.B.Y. Cox and A.S. Somers Cocks as masters from the 1890s until the 1920s.[51] These last were both outstanding players for Barbados, and both subscribed fully to the English public school ideology. Harrison College poured out great cricketers, schooled also in the game's social code, such as H.B.G. Austin and George Challenor who made his first tour for the West Indies to England in 1906 while still a student. Victories by Harrison and, indeed, by Lodge over teams of grown men were greeted warmly and as evidence of the power of cricket as a social training ground for adult life.[52] School cricket was central to a cultural reproduction system which maintained and boosted the code of respectability and which in itself was a major contributor to the classed nature of Barbadian cricket organization.[53]

Stribling's initial murder suspect attached great importance to being 'a gentleman in a gentlemen's cricket club'. As a Barbadian professional he could never have been a gentleman, of course, but the important point is that he *thought* he was. Barbadian cricket created the illusion of social equality but, at the same time, preserved rigid and complex social distinctions. Sport in most cultures has been reckoned to ignore normal class and status divisions but in reality it cannot; like any social institution it is constituted by the attitudes and prejudices of its host community.[54] Caribbean cricket was certainly not egalitarian, being predicated on class and colour.

Few passages in *Beyond A Boundary* are so rich as James' description of his dilemma about which Port-of-Spain club he might join after leaving Queen's Royal College, the Trinidad equivalent of Harrison College. As an educated dark-skinned man, he could not or would not join certain clubs: Queen's Park catered mainly for wealthy whites and established mulattos, Shamrock had an almost exclusively white Catholic clientele, and Stingo was a lower-class black preserve. His choice lay between Shannon, a black lower-middle-class club, and Maple, a brownskin middle-class club. Such distinctions were repeated throughout the English-speaking Caribbean, varied only by the specifics of the population mix – Chinese and East Indian teams, even competitions, appeared in British Guiana and Trinidad. Class and status wrought similar distinctions within all cricket competitions.

Around 1900 in British Guiana the Georgetown Cricket Club was dominated by the white élite drawn mostly from the top civil service echelons; the expensive $10 entrance fee and $10 annual subscription helped maintain its exclusiveness. The British Guiana Cricket Club attracted more junior civil servants and merchant clerks who could afford the 10 shillings entrance fee and $5 annual subscription. The British Guiana Churchmen's Union Cricket Club (evidence of the 'muscular Christian' strain in cricket at the time) drew upon clerics and school teachers for the most part although J.T. Chung, one of its captains, was a clerk on a Chinese-owned estate. Up country in Berbice the County Cricket Club was run by the regional Commissioner of Taxation assisted by an auctioneer, a clerk, a copyist and an accountant.[55] Over in Jamaica the Kingston Cricket Club was for the colonial white élite; the Kensington Cricket Club began in 1879 among lower-status school products; the Melbourne Club was formed with an educated black leavening early in the 1890s; Lucas, a poor black club, emerged after the 1895 English tour and took its name from Slade-Lucas, the touring captain.[56] These clubs preserved and boosted their social limits well after 1914 and so it was in Barbados where there were eight competition teams for much of the period under review.[57]

Harrison College and Lodge never won a championship before 1914 and were often chopping boards for more powerful sides. Although there were no immutable rules, Harrison boys leaving school generally moved to Wanderers, those from Lodge to Pickwick. There were exceptions, of course, generally based upon social or occupational position, geographical location, or skin colour – until 1915 black school-leavers had Spartan as their only club choice. Then there was the Garrison, a team drawn from the white British troops stationed in Barbados until the early twentieth century. Leeward and Windward were two 'country' clubs, drawing their members from the white planting and managerial community at opposite ends of the island. Their matches against each other were traditionally the occasions for great celebration, the reaffirmation of the planting faith, so their high teas were legendary.[58] The social nature of their matches often meant scores were never relayed to Bridgetown, but the purity of their play and devotion to the higher cricket traditions were admired widely. The remaining clubs were in reality the 'big three': Wanderers, Pickwick and Spartan.

From its 1877 foundation until 1914 Wanderers was the most prestigious club in Barbados. During that time it had strong planting connections: among its captains were D.C.A. Ince whose family was connected with estates in the Christ Church parish; E.A. Hinkson who inherited the 340-acre Coconut Hall in St George and who married a daughter of Horace Deighton;[59] and Kenneth Mason whose family also held property in St George, the district for which he became Parochial Treasurer in 1903. Many other planters were prominent Wanderers players. Then there were the influential and respectable merchants: da Costa, father and son; Robert Arthur who inherited his father's business; the Collymores, the Austins and the Challenors. Other players were sometimes senior employees of these firms. Respectable English arrivals invariably gravitated to Wanderers. William Bowring arrived in the late 1890s, was a prominent player for Barbados and West Indies, married a da Costa, became a director of the firm and first President of the Barbados Cricket Association established in the 1930s. M.L. Horne arrived as a bank clerk and became a fast bowling hero early in the twentieth century, taking 8 for 39 against the 1902 English tourists. A.S. Bryden junior joined the club after being educated at Tonbridge School in England, and later married the daughter of club President and legal luminary W.K. Chandler. Numerous senior civil servants were members. F.B. Smith entered the service in 1859 to become Inspector-General of Prisons and Provost Marshall among other posts; he batted through the Barbados innings in the first inter-colonial match in 1865, and influenced Wanderers considerably as secretary. W.H. 'Billy' Allder, the son of a cleric, succeeded Smith as secretary and was chief registers clerk in the Colonial Secretary's Office. Beyond that were lawyers like E.C. Jackman and C.E. Yearwood (who married J. Gardiner Austin's daughter), medical men like Dr John Hutson and clergymen like the lively Reverend C. King-Gill. Wanderers, then, drew on the leading quarters of Barbadian society throughout the period.

Pickwick rivalled Wanderers on the field but not in the social stakes where it lacked the same prestigious profile. There were few, if any, major planters although the Goodman brothers came from a family in the industry, and two of them (Clifford and Percy) married into planting interests while Aubrey, the lawyer, became a landholder in St. John. The merchants were generally less important or newer to wealth than those in Wanderers. J.R. Bancroft, for example, worked for Challenors after leaving Harrison College, later became a partner in a smaller firm then moved into directorships as well as property.[60] One of his sons married a da Costa and another, a Barbados Scholar, was wicketkeeper for the 1906 West Indies tour of England. J.C. Hoad, an early secretary, ran a provisions business and the Hoad name was prominent in Pickwick circles until the 1960s. J.H. Emptage, an accountant, was an early secretary, and others of the same name and similar occupations followed. Lawyers were few, and G.O.D. Walton, who struggled for years to receive a permanent position on the bench, finally received one on the island of St Kitts, and was later knighted for his services to the Caribbean judiciary, was the most successful. There were numerous lower-level clerks both in business and the civil service, and some teachers: Percy Goodman became principal of Christ Church Foundation School, while L.A. Walcott, son of an accountant, an Island Scholar and Barbados player, was games master at both Harrison and Lodge between the wars and after. The touch of respectability was there in Pickwick but did not match that of Wanderers as the case of that magnificent player Tim Tarilton suggests. His father was Poor Law Inspector in St Johns and sent his boys to Lodge. Tim joined the Customs service, married the daughter of a veterinary surgeon and, in 1919, was appointed Parochial Treasurer of St James. He was clearly great friends with George Challenor (with whom he set many opening partnership records for Barbados and West Indies) and H.B.G. Austin (his captain in those teams). Just as clearly, he was from a different social world from the one inhabited by his playing colleagues, and in Barbados such different social circles very rarely mixed. In that difference, perhaps, originated the fierce Pickwick competitive spirit and club loyalty which appeared during 1895, for example, when some members refused to play for Barbados because other clubmates were not selected.[61]

The Spartan spirit lay in its representation of the emerging urban coloured middle and upper-middle class (it is said that lightness of colour was an important admission criterion).[62] The club joined the Cup competition in 1893 playing on the ground in Belleville, the new respectable suburb. Sir Conrad Reeves, son of a white father, was its first President and a long serving Vice-President was J. Challenor Lynch. Lynch was a coloured, non-practising lawyer who inherited his father's merchant business and who, like his father, became a prominent politician. The firm was deeply involved in plantation transfer. As early as 1891, for example, Lynch bought in Chancery the major 454-acre Pine Estate, and he lived on another excellent property, Friendship. Then, the Rudder family was prominent in both playing and administration: Stephen and George, for example, were the sons of W.A. Rudder who was in planting as early as the 1860s and who gave evidence against Federation rioters. Stephen (who played for Barbados) reached senior civil service positions while his brother became Poor Law Inspector in St Thomas. There were medical men like Dr C.W. St John, lawyers like H.M. Cummins and G.H. Corbin. C.A. (Johnny) Browne was a well-to-do jeweller, brother of P.N. Browne who established the business, then moved to British Guiana where he moved into politics. He was joined there, at the end of this period, by another brother, C.R. (Snuffie) Browne who trained as a lawyer in England. The Brownes' father was among the first non-whites to enter Bridgetown politics. Snuffie established an excellent record as a West Indian all-rounder while Johnny played for Barbados as a stylish, hard-hitting batsman. A prominent founding player, captain and club power was Graham Trent Cumberbatch, son of a successful bootmaker. He took a degree through Codrington College and became Assistant Inspector of Schools but, like others in his social position, found progress difficult. When in 1901 he applied for the position of Inspector of Schools in the Leeward Islands, the Governor thought it necessary to inform the Colonial Office that he was 'a man of colour' and, by implication, inappropriate for the post.[63] Two general reactions to such discrimination emerged in Spartan. On the one hand, men like Cumberbatch and Cummins became ardent enthusiasts of the cricket ideology, attempting to share the cultural values of the whites with whom they competed both in cricket and in society. On the other hand, they developed a strong desire to win, to beat the representatives of those who displayed the prejudice. The essential paradox in this dual position is clear. While trying to emulate the ruling cricket and social values, Spartan members had also to deal with the inequalities contained in those ruling values. On the whole, Spartan men resolved to accept the inequalities, an excellent demonstration of Gramsci's theory of hegemony.

Given this general background, then, it was scarcely surprising that some Barbados club fixtures became exceptionally competitive, attracting large and partisan crowds. The most famous series was between Wanderers and Pickwick, usually the talk of the town for weeks beforehand.[64] But the inter-school matches always had an extra edge to them, Spartan singled out Harrison for particular attention,[65] and Pickwick versus Spartan also attracted notoriety. While the Barbados Cricket Committee for the period (manned by lawyers, merchants, senior civil servants but scarcely a coloured or black representative) promoted the idea of social unity through cricket, in reality there developed a clearly defined and stoutly defended social hierarchy based on caste and colour, and which could not always be hidden. As early as 1891, one commentator asked that the Barbados team be selected on grounds other than caste or creed.[66] But it was the 'Fitzy Lilly' affair of 1899 and after which showed how closely guarded social rankings became.

Fitzy Lilly, real name Fitz Hinds, was a painter by trade but more importantly, during the early and mid-1890s he was a prominent net bowler in the little layered world of the Bridgetown cricket clubs.[67]

From the first appearance of organized clubs, a large number of lower-class blacks became skilled cricketers, especially bowlers, simply by giving practice to competition players. All clubs had them and Fitz Hinds was attached officially to Pickwick. By the late 1890s, however, he had decided upon a club playing rather than service career so dropped out of Pickwick for a year; he put up for membership at Spartan where Cumberbatch and Cummins were his chief supporters. At the first ballot he was rejected by four out of the 16 voters, his inferior social status clearly counting against him. Only some smart committee work won him membership a few weeks later.[68]

By then, of course, the matter had reached the attention of other clubs, particularly Pickwick with its keen sense of social position (and where Hinds had been a servant) and Windward which contained a particularly conservative planter element.[69] Leading players in both teams refused to meet Hinds on the field so that the man suffered on all sides – one report had it that he was snubbed at Spartan practice sessions by those opposed to his membership.[70] His reply against Pickwick (where Clifford Goodman led a fight to recognize the ex-pro) was to topscore with 24, then take 8 wickets for 31 to give Spartan a rare victory. With the two rebel clubs refusing to field strong teams, Spartan swept to its first Cup win and Hinds did so well as to tour England with the 1900 West Indies side (one Pickwick player refused to go because Hinds was selected).[71] He remained a prominent Barbados player until he shifted to the United States in 1905 where he continued to play well, appearing against an Australian team there in 1913.[72] His was a stirring achievement under intense pressure which arose from the social layering of Barbadian cricket, itself produced by the island's sugar culture which allocated all members of the community a rank in its elaborately defined production hierarchy. It is important to note that colour was not the sole, or even the important, issue in the Hinds case; it was social position, as his rough treatment inside Spartan indicated. Barbadian cricket ideology and its imperial model demanded a minimum status for admission to its organized ranks. For that reason, the other importance of Fitz Hinds is that he indicated the presence, both formal and informal, of a substantial cricketing body outside Barbados Cricket Committee auspices. This sub-culture, as it might be called, is important because its fortunes, until 1914, indicate indirectly the power and success of the élite traditions and practices.

Stribling erred again in having his professional cricketer and murder suspect come from the poor white class; in Barbados such players were always black (the demographic variations of Trinidad and British Guiana meant that there a few emerged from the East Indian and Chinese communities).[73] But Stribling was right in his depiction of the professional's dependent social and economic position. By 1895 in Barbados there were at least 14 ground staff attached officially to six of the competition clubs, and earlier than that Wanderers and Pickwick alone supplied enough talented professionals to test an Island XI practising for an inter-colonial.[74] Many professionals had long club residencies: William Shepherd was with Spartan from the start until well into the twentieth century (in the off-season he frequently took his talents to New York), while Germain was a long stayer at Pickwick as was Oliver Layne at Wanderers. Their duties were twofold: to bowl to club members in the nets, and to prepare the characteristically shiny Barbados pitches which often reflected the players as in a mirror. For this they were paid a retainer, members frequently gave tips for their services, and they earned additional income from umpiring. It was a steady if hard life. Joe Benn and Tanny Archer, good enough players to be selected against touring teams, both died from tuberculosis.[75] With people like 'Chingham alias Fonix alias Phinx' it is difficult to sketch a background, but William Francis Jones and Oliver Layne are a little different.

Jones was born in Bridgetown during 1875 to Rosetta Jones of Constitution Road, the 17-year-old daughter of Elizabeth Jones from Codrington Hill.[76] In neither case is it possible to trace a father, but clearly either grandmother or mother, possibly both, had shifted to Bridgetown as part of the urban drift. William Jones was baptized in St Michael's Cathedral so it seems likely he lived close to the Bay ground where he was on the staff by at least 1895. Oliver Layne was born almost exactly a year after Jones. His father, Richard Layne, listed himself as a porter (a handcart operator) of Britton's Hill who had married Margaret Wharton in St Michael's Cathedral late in 1872. Like Jones, Layne was on the Wanderers staff by at least 1895, but by 1909 he was playing in Trinidad where more open attitudes provided Barbadian professionals with wider playing opportunities. Like Fitz Hinds, Layne then shifted to the United States where he played in the New York leagues before dying in the early 1930s. Archie Cumberpatch, among the earliest in a long string of famous West Indian fast bowlers, was another who left Barbados for Trinidad in the late 1890s and immediately attracted the attention of English touring players. Like Cumberpatch, Jones and Layne were excellent players, the latter taking 6 for 8 against a good Spartan side in 1900 and taking 57 wickets on the 1906 West Indies tour of England. By the turn of the century, opportunities for such men had broadened a little in

Barbados because of the remarkable William Shepherd.

Building on his career at Spartan, Shepherd established a regular team of ground staff which was soon playing club teams in practices and providing matches for the Barbados team before inter-colonial tournaments. He expanded professional opportunities in 1900 by taking the team on tour to Trinidad, followed that with a British Guiana tour in 1904 then, during 1909, organized an inter-island team of professionals which he led through both Trinidad and British Guiana.[77] Shepherd was no ordinary net player, then, nor was his an ordinary skill. Against a Barbados XI in 1899 he took 8 for 45, and in 1900 against Pickwick made 54 runs out of 108 and then took 8 for 24. His playing and entrepreneurial skills helped professionals to be included in Barbados teams against touring sides from 1902 onwards. But even his efforts and the rapidly growing skills of his fellow professionals could not break the élite social strictures of the Barbados Cup competition – calls for the inclusion of professionals were consistently rejected on social rather than playing grounds, the argument being that club members and friends went to cricket to meet others of their rank and station.[78] This attitude meant Shepherd could enter his talented Fenwick team in what became known as the Frame Food competition but not in Cup fixtures.

When he was organizing Fenwick Shepherd informed at least one newspaper that a number of his players were *not* professional.[79] Some of those he mentioned were listed frequently as attached ground staff, which raises the distinct possibility that players referred to casually as 'professionals' may have been employed elsewhere. That is, they may have been working-class players improving their skills (and, incidentally, their incomes) by meeting Cup players in the nets. If that was even partly true, then there are at least two further interesting possibilities: the existence of a strong working-class cricket tradition in Barbados, and a growing rather than declining desire by the sugar and cricket élite to remain socially separate from such a tradition. Some scattered evidence supports these possibilities.

First, there was the immediate numerical strength of the Frame Food sponsored competition established in 1902 by Frederick Martinez, a man known throughout the Caribbean and Latin America as the 'prince of Commercial Travellers'. Drawing its name from a milk-based food product popular with the working classes, the competition aimed at boosting cricket among those classes. Sponsored by a low-status commercial man and with little support from respectable quarters, the competition stood little chance of success. It collapsed within four seasons although its cup remained at stake for matches between working-class clubs until the 1920s.

Thirteen teams competed in the first year and included Police, Railways, West India Regiment (black troops), Volunteers, and Fenwick which won the first two competitions.[80] Bankers entered a team later.[81] This was a clearly occupational and frequently artisan-based organization coming from earlier, sporadic holiday fixtures such as Printers versus Tailors.[82] Just as clearly, that number of teams and the administration required to organise them would not have developed instantaneously just because Martinez donated a trophy. This cricket must have been in existence for some time, its claims and growing strengths studiously ignored by the Barbados Cricket Committee on social rather than cricket criteria. For that reason, evidence relating to such cricket is rare, the bulk of cricket records being maintained by 'respectable' authorities with a sense of social mission. Then, the connection between this and Shepherd's apparently 'professional' team is provided by S.A. Merritt, a regular Fenwick player. He was appointed Secretary to the committee which ran the ambitious Frame Food competition. As he was in the company of an engineer, a company clerk and a freeholding wheelwright, he must surely have possessed skills other than those of a net bowler.[83] Fenwick, then, might have been as much the expression of frustrated working-class cricket as it was that of the frustrated professionals. But the Barbados Cricket Committee and the social system which it represented rejected both thrusts resolutely, though it did nominate William Bowring as its observer on the short-lived Frame Food Committee. For the most part, respectable society and cricket kept its distance seeing the 'excitement' caused by working-class matches as symptomatic of the social conditions it was trying to overcome.[84] But that was not the only alternative faced by organized cricket, for there remained the even more profane game played at large and in public thoroughfares.

In 1891 the 'stave bat and primitive ball of native manufacture' were taken for granted in a debate on public recreation facilities.[85] After the 1895 tour by an English team men and boys were said to be playing in every alley and field.[86] And in 1905 the English tourists were delighted by the fielding of the little boys who turned up to 'assist' them at the Bay ground.[87] Lower-class enthusiasm and élite concern for the game sometimes met in the courts, belying the view that in cricket all other inequalities were forgotten. Nathaniel Hutchinson must have tired of the 'primitive ball' because he received 14 days with hard labour 'for the larceny of a Cricket-ball' from Wanderers' pavilion, an illustrative meeting of very different cultures with quite separate cricketing and social objectives.[88] In 1899 a woman charged that two men playing cricket hit her first with the ball then with their fists; they counter-charged that she had detained the ball unlawfully! One was fined two shillings plus six shillings costs in lieu of seven days with hard labour, not at all the image sought by the promoters of organized cricket (it is too much to think that the Archie Matterson involved was the same man who played for a Ground staff XI against an Island team that year and who became treasurer to the Frame Food Committee!).[89] However physically close to organized, respectable cricket this activity might have been, it was a vast social distance from the glitter of Wanderers at the Bay, confirming the rigid and complex social divisions which contradicted assertions that the whole society met in cricket and raising the question, naturally, of how successful the promoters of orthodox cricket ideology were in penetrating Barbadian culture and influencing the masses towards accepting their social standing. Like other writers, Stribling depicts Barbados as being more obsessed by the game and its Englishness than any other Caribbean community. But before 1914, at least, there were really three varying responses.

Outright acceptance of cricket and its English cultural provenance was demonstrated best by 'Britannia Bill', the Union Jack-carrying black who met the 1897 English team, watched their matches and followed them about praising their skills, character, country and monarch – 'Old England Forever' was his constant cry.[90] In 1905 a black spectator kissed the arm of an English bowling star.[91] Similar sentiments were expressed to most pre-1914 touring teams whose members were greeted warmly in their Bridgetown travels, especially 'Steady Stoddart' as he was dubbed, the most accomplished player to visit Barbados before the First World War.[92] There is the thought that some of this might have been directed against the local élite; a modified form of colonial resistance stemming from a natural warmth towards representatives of the Queen and country which had formally ended slavery. However, similar crowds turned out to watch the nearly all-white inter-colonial and the socially exclusive local club matches whose players received similarly approving treatment. An identification with the cause and philosophy of Barbadian cricket undoubtedly existed and 'the great unwashed' displayed a close knowledge of the game from very early on. Within half an hour of a 1900 Barbados victory being telegraphed from Trinidad, 'idle labourers' and others were discussing and celebrating it animatedly.[93]

Modified acceptance of the pure model was shown by the black who in 1895 was 'sanguine that we may yet propel our flag among the nations as the Colony which has humbled its Mother'.[94] The colonial élite could be happy with that, of course, because it demonstrated an identification with Barbadian interests. But it also indicated the beginnings of a local modification which did not meet with wholehearted acceptance from the authorities – crowd participation. If the crowds in the 'ring' (the cheapest sections of the grounds) were knowledgeable, they were also noisy and unhindered by the social reserve thought proper among cricket watchers reared on English precedent. Barbadian crowds offered a constant stream of advice to batsmen, bowlers, fielders and captains. One 1905 English player heard his bowlers urged to 'bowl him down' and the Barbados batsmen cautioned to 'watch dat man at square leg'.[95] And it was during this period that the ring developed its passion and encouragement for spectacular fast bowling. In 1900 larger crowds than normal followed the Wanderers team captained by E.A. Hinkson, lured by the *white* fast bowler M.L. Horne. They would applaud Horne as he arrived at the ground, then wait in the expectation that 'now Hinkson gwine tek he Horne an butt dem down'.[96] Serious crowd demonstrations were rare (perhaps suggesting the strength of the behaviour code) but it was always with nervous self-assurance that good crowd temper was mentioned by the representatives of respectable society; and at club fixtures unrest was quelled immediately.[97] And there was always official embarrassment when visiting teams (the sensitive English particularly) had difficulty accepting local crowd performances.[98] Percy Goodman in 1909, for example, assured Trinidadians that a hissing incident over an umpiring decision had arisen in an unknowledgable section of the crowd and was quite isolated and unusual (some thought it started in the élite pavilion!).[99] Such animated behaviour, however, had come to stay as perhaps the major local modification to imperial cricket ideology.

There are few explicit examples of direct rejection of or opposition to that cricket playing and behavioural code constructed by the Barbadian élite. A match at Lodge was interrupted by Federation rioters in 1876, but their objective was the adjoining Guinea estate rather than the players or their field.[100] It is also difficult to make out a case of premediatated colonial resistance by the two boys who in 1897 tore planks from the Pickwick pavilion, or by those who arrived at the Wanderers ground during an 1899 match to deride with a local form of taunt: 'Well played, Sop Biscuits'.[101] In each case, though, the culprits were dealt with promptly and severely for having brought disrepute to the game and its social mores.

The classic resistance weapon of rumour touched cricket occasionally. In 1897 relatives of T.W. Roberts were informed by 'bush telegraph' that he had broken his neck while playing for Barbados against the English tourists. It was incorrect, but the origins and purpose of the story remain intriguing.[102] An even more fascinating and direct act against cricket interests occurred during the 1895 English visit when an unknown hand turned on a tap and flooded the pitch on the eve of the first day. Local sources put it down to an 'inexplicable' bit of 'devilry' by a 'wicked urchin', while the visiting captain thought it the work of a 'native' (presumably lower-class) annoyed at having to pay an entrance fee for the first time.[103] He might well have reworked that as an expression of discontent at the exclusive nature of Barbadian cricket expressed in the size of the fee. That was an extremely tough year economically, and crowds of rioters later looted fields in Clermont, Boscobelle and elsewhere throughout the island while singing songs of Federation, invoking memories of the last major occasion on which the power of the sugar hierarchy had been challenged.[104]

There is little doubt that C.L.R. James was correct in identifying cricket as a major bulwark against social and political change in the English-speaking Caribbean. In Barbados and the other colonies, until 1914 at least, the colonial élites established a cultural primacy through cricket as much as through economic power and political position. As Gramsci suggests, these élites established and maintained their positions by determining that their values, traditions and standards be accepted by the populace at large as the cultural programme most appropriate to the community, even though the bulk of that community had no access to the institutions through which the programme was inculcated. One principal reason for the important place held by cricket was the significance of the game to the imperial power itself. Cricket and its associated moral code were regarded in two specific lights by the Caribbean élites: as a means of forging a close cultural bond with the imperial power, and as a means of establishing themselves as the arbiters and agents of that imperial philosophy within their own social and political environments. Consequently, agencies such as the church and the élite schools became as important in the Caribbean as they had been at home in fostering the skills and social traditions which carried in them the imperial messages of cricket. The consistent recruitment of religious and educational personnel from Great Britain was just the most obvious indicator of this cultural reproduction until the eve of the First World War.

For the most part, the colonial élites carried on this process unhindered, controlling those agencies of social reproduction identified by Pierre Bourdieu as being central in the creation of hegemonic cultural values. Cricketers outside the élite groups made little headway before 1914. A number of professionals emerged from the working classes, but they were the most bound to the ruling-class view of the game because their livelihood depended upon pleasing their masters. Men like William Shepherd were extremely careful lest their dress, demeanour and deportment should displease their superiors. Then, some non-élite competitions began before 1914, such as the Chinese and East Indian leagues in Trinidad and British Guiana as well as the Frame Food competition for lower-class blacks in Barbados. Most failed, with the provision of regular cricket for such social groups coming well after 1918. But before their failure, most of these non-élite competitions had demonstrated the power of the élite position by attempting to impose upon non-élite groups the moral and ethical codes derived from the imperial cricket model. Indeed, the failure of the alternative competitions was attributed as much to an inability to maintain these codes as to the antagonism of the controllers of élite cricket.

As for popular support, cricket was a positive mania in most areas of the British Caribbean by 1914, the feats of the cricketing and social élite cheered on by the masses who shared a belief in the social importance of the game if not the access to its playing arenas. The controllers of cricket were quick to point out that Caribbean crowds appreciated fine play (irrespective of teams), sportsmanship, dash, courage and temperament the essential qualities of the imperial cricket ideology. Although the crowds were noisy and demonstrative, the élites interpreted this as only to be expected, coming as it did from the descendants of 'excitable' Africans and mixed-blood populations. Indeed, it was considered a testament to the social power of cricket that such crowds were not even more noisy and demonstrative. In all aspects of Caribbean cricket, then, the colonial élites' hold, organization and philosophy were deep-seated, with ramifications into political, commercial and cultural life, as C.L.R. James and others like him were to discover.

By 1914 cricket in the Caribbean, as exemplified by the Barbadian case, was not a monolithic social agency in which all classes and colour groups met uncaring of the differences encountered in other sections of civic life. The game was played at all levels of society, clearly enough, but there were distinct and accepted areas of activity for them as potential reformers discovered. Within each level, too, a social coding system saw players, administrators and often spectators *by choice* move to their appropriate locations. Fitz Hinds proved the power of that. Immediately he entered Spartan the Barbados Cricket Committee enacted a new rule to ensure no repeat of the incident.[105] These elaborate social springs were situated far below the surface of Caribbean colonial life, their flavour determined by the specific local cultural mix of demographics, economy and patterns of prestige. Far from waning by 1914, this complex cultural power structure was still growing while other colonial bonds were showing signs of deterioration. This was demonstrated best in Barbados, appropriately enough, in the 1915 secession from Spartan which created the less socially prestigious Empire club, directed by the legendary Herman Griffith whose wonderful fast bowling and quicksilver temperament presented for the next generation a combined challenge to both cricket authority and the island culture which had produced it.[106]

NOTES

The research for this paper was made possible by a Professional Experience Programme leave from the Canberra College of Advanced Education, as well as by the kindness and encouragement of the History Department, University of the West Indies, Cave Hill, Bridgetown, Barbados which elected me an honorary Visiting Fellow for 1985. Thanks are due also to the staff of the Barbados Archives, to those who participated in the seminar on the original version of the paper, to Mrs Ida Kidney, and to Mr W.F. 'Ben' Hoyos. Particular thanks to Professor Woodville Marshall and Dr John Mayo.

Abbreviations
BCA *Barbados Cricketer's Annual*, published annually between 1894 and 1914, and compiled by J. Wynfred Gibbons.
BA Barbados Archives.
Note on newspapers
Where no page reference is cited, pagination was not employed in the original. For convenience I have dropped the word 'Barbados' from the *Barbados Argicultural Reporter*, *Barbados Bulletin*, *Barbados Globe*, *Barbados Advocate* and *Barbados Herald*.

1. T.S. Stribling, 'Cricket' in *Clues of the Caribbees: Being Certain Criminal Investigation of Henry Poggioli, PhD* (New York, 1977 edn).
2. For the method Michel Foucault: *Order of Things: Archaeology of the Human Sciences* (London, 1974). A good descriptive rather than analytical history of the island's cricket is Bruce Hamilton, *Cricket in Barbados* (Bridgetown, 1947). He shows the early and inordinate hold taken by cricket upon the population.
3. Pelham Warner, *Cricket in Many Climes* (London, 1900); Lord Hawke, *Recollections and Reminiscences* (London, 1924).
4. C. Wright Mills, *The Sociological Imagination* (Harmondsworth, 1973 edn.), especially Ch. 8.
5. C.L.R. James, *Beyond A Boundary* (London, 1963).
6. H. Orlando Patterson, 'The Ritual of Cricket', *Jamaica Journal*, 3 (1969); Maurice St Pierre, 'West Indian Cricket – a Socio-Historical Appraisal', *Caribbean Quarterly*, 19, 2 and 3 (1973). Frank E. Manning's title 'Celebrating Cricket: the Symbolic Construction of Caribbean Politics', *American Ethnologist*, 8 (1981) is a little misleading here in that it deals with an area rather outside the Caribbean cricket culture proper. See also Vince Reid, 'Thoughts on England v. the West Indies, Edgbaston, 1973: Or the People of the Caribbean Still Need to be Emancipated From One Type of Domination or Another', *New Community*, II (1973).
7. See the hopes of creating a West Indies touring team early in the 1890s, *Globe*, 20 February 1893. West Indians and British imperialists alike maintained the ideal of cricket and federation beyond the First War; L.S. Smith, *West Indies Cricket History and Cricket Tours to England* (Port of Spain, 1922), pp. 6, 73, 127.
8. This section based on: *The British Guiana Directory and Almanack* (Georgetown,

1906); *The British Guiana Handbook* (Georgetown, 1922); T.B. Jackson (ed.), *The Book of Trinidad* (Port of Spain, 1904); *Trinidad and Tobago Yearbook* (Port of Spain, 1938); *Handbook of Jamaica*, (Kingston, 1905). Smaller islands are not covered here, but most if not all played cricket; for example, J.H.D. Osborne, 'Cricket in St Lucia', *BCA* 1901–02, 63 ff.

9. For example, J.I.M. Hewitt, 'West Indies in Big Cricket', *Silver Jubilee Magazine* (Bridgetown, 1935), 26; L.S. Smith (ed.), op. cit., p. 7.

10. See, for example, the quarrel over the 1900 team involving W. Bowring *BCA* 1899–1900, 136–137; *Agricultural Reporter*, 27 January 1900, 2–3.

11. *Queen's Park Cricket Club Diamond Jubilee* (Port of Spain, 1956), p. 91; W.A.S. Hardy (ed.), *They Live For Cricket* (London, 1950), pp. 12, 21.

12. *BCA*, 1895–6, 15.

13. Statistics from *Barbados Blue Book* (Bridgetown, 1900).

14. For the brief history of Barbados: F.A. Hoyos, *Barbados: a History From the Amerindians to Independence* (London, 1978); George Hunte, *Barbados* (London, 1974).

15. *Globe*, 16 October 1893. Also *West Indian*, 7 Jan. 1876 and 19 January 1877.

16. A description of the working of the Act is in Governor to Secretary of State, 27 Sept. 1881, Confidential, COL 2/1/26, BA. For an interpretation of post-emancipation industrial relations, B.M. Taylor, 'Black Labor and White Power in Post-Emancipation Barbados: a Study of Changing Relationships', *Current Bibliography on African Affairs*, 6 (1973).

17. See John Wickham, *Nation*, 10 March 1985, 4, for a discussion of the sugar culture.

18. For example, *Agricultural Reporter*, 4 Sept. 1877, and 9 Oct. 1877.

19. *BCA* 1894–95, 16–28.

20. Ibid., 1896–97, 171–2.

21. *Advocate*, 13 Feb. 1905, 6–7.

22. *BCA*, 1897–8, 30 and 1908–09, 141–2.

23. *Bulletin*, 17 Sept. 1897, 7–8. And, of course, 'What is death to cricket is life to the planting interest', *Bulletin*, 10 Sept. 1899, 9.

24. Paragraph based upon analysis of Voters Lists, BA, for the period 1880–1914.

25. For some indications on the economy: Governor to Secretary of State, 26 July 1884, Confidential, COL 2/1/28; BA; *Globe*, 21 Sept. 1893; *West Indian Royal Commission: Barbados* (Bridgetown, 1877); Governor to Secretary of State, 15 Jan. 1902, no. 12, COL 2/1/44, BA. On merchant loans and purchases see, for example: *Agricultural Reporter*, 17 Feb. 1900, 1; 20 Jan. 1900, 3; 10 March 1900, 3; *Globe*, 19 Oct. 1893; *Advocate*, 14 Jan. 1905, 5. Other 'rescue' legislation included the Plantations-in-Aid Act and the Sugar Industry Agricultural Bank: see Governor to Secretary of State, 20 August 1902, No. 177, COL 2/1/45, and Governor to Secretary of State, 4 March 1907, No. 30, COL 2/1/47, both BA.

26. *Advocate*, 11 Jan. 1908, 6 and 4 Jan. 1909, 6. The 1909 case concerned F.M. Alleyne who had owned Kensington estate from which Kensington Oval, home of Pickwick and Barbadian cricket had emerged; Hiralal T. Bajnath (compiler), *West Indies Test Cricketer* (Port of Spain, 1965), p. 88.

27. Unless specifically noted otherwise, biographical details for this paper have been accumulated from sources such as: Voters Lists; Gabriel Anciaux, *The Barbados Business and General Directory 1887* (Bridgetown, 1887); S.J. Fraser, *The Barbados Diamond Jubilee Directory and General West Indian Advertiser* (Bridgetown, 1898); *Barbados: Historical, Descriptive and Commercial* (Bridgetown, 1911); E. Goulbourn Sinckler, *The Barbados Handbook* (London, 1913); *Leverick's Directory of Barbados* (Bridgetown, 1921); *Barbados Yearbook and Who's Who* (Bridgetown, 1935); *BCA* 1894–1914; newspapers.

28. For very different views on the origins, nature and purpose of the riots: James Pope-Hennessy, *Verandah: Some Episodes in the Crown Colonies, 1867–1889* (London, 1964), Book V; Bruce Hamilton, *Barbados and the Confederation Question, 1871–1885* (London, 1956); George A.V. Belle, 'A Study in the Political Economy of Barbados, 1876–1971' (Mona, 1972), and 'The Abortive Revolution of 1876 in Barbados' (Cave Hill, 1981).

29. For example, *West Indian*, 9 June 1876.
30. *Agricultural Reporter*, 2 March 1877; *West Indian*, 16 March 1877 and 27 March 1877; evidence in *Report of the Commission on Poor Relief, 1875–1877* (Bridgetown, 1878); *Times*, 12 April 1882; Anciaux, op. cit., p. vi; *Agricultural Reporter*, 12 September 1893; *Globe*, 12 April 1894; J. Gardiner Austin evidence before *West Indian Royal Commission: Barbados*, op. cit., p. 15; *Bulletin*, 21 September 1899, p. 8; *Agricultural Reporter*, 8 March 1900, pp. 2–3; *Advocate*, 6 March 1908, p. 5 and 22 March 1909, p. 5.
31. Paragraph based on: *Agricultural Reporter*, 10 August 1877; Governor to Secretary of State, 15 May 1880, no. 70, COL 2/1/26, BA; *Globe*, 22 December 1892; Governor to Secretary of State, 28 June 1899, no. 154, COL 2/1/44, BA; *Agricultural Reporter*, 5 April 1900, p. 4; *Advocate*, 22 March 1909, p. 8.
32. The theory and practice for this general point lies with the field of cultural studies, especially the early work of the Centre for Contemporary Cultural Studies at Birmingham University. For some of the dimensions: Stuart Hall, 'Cultural Studies: Two Paradigms', *Media, Culture and Society*, 2 (1980); Raymond Williams, *Problems in Materialism and Culture* (London, 1980) and *Culture* (London, 1981).
33. *Wanderers Cricket Club Centenerary, 1877–1977* (Bridgetown, 1977), p. 27; *West Indian*, 9 May 1876; *Agricultural Reporter*, 4 Sept. 1877; Governor to Secretary of State, 9 September 1876, No. 200, COL 2/1/25, BA.
34. *Barbados: Historical, Descriptive and Commercial*, op. cit., p. 53.
35. *Times*, 7 Jan. 1882.
36. For an introduction to Gramsci see James Joll, *Gramsci* (London, 1977) and Joseph V. Femia, *Gramsci's Political Thought: Hegemony, Consciousness and the Revolutionary Process* (Oxford, 1981); for a sample of the work, Antonio Gramsci; (translated by Louis Marks), *The Modern Prince and Other Writings* (New York, 1975).
37. Jackson, op. cit., p. 148; *Handbook of Jamaica*.
38. For a flawed analytical version of the English story, see Christopher Brookes, *English Cricket: the Game and Its Players* (London, 1978), and for a comparative case of its colonial influence, Richard Cashman, *Patrons, Players and the Crowd: the Phenomenon of Indian Cricket* (Bombay, 1980).
39. See *Globe* for Oct. 1894; *Bulletin*, 9 Sept. 1897, 12–13; *Advocate*, 13 Feb. 1902, p. 7 and 16 Jan. 1909, p. 7. In addition, the 'symbolical inscription of the great game' on the Intercolonial Cup was said to depict Dr W.G. Grace, *Globe*, 28 Dec. 1893.
40. See the editorial, *Herald*, 28 Jan. 1895. This concern was widespread throughout the Empire; for example, W.F. Mandle, 'Cricket and the Rise of Australian Nationalism', *Journal of the Royal Australian Historical Society* (1973).
41. C.P. Bowen, *English Cricketers in the West Indies* (Bridgetown, 1895), p. 6; *Herald*, 25 April 1895.
42. *Agricultural Reporter*, 1 May 1899, 3–4.
43. *Belfast Telegraph*, 26 July 1923, Cricket Scrapbook 1923, BA. An even greater emphasis on the role of cricket in bringing English social life to the Caribbean was made in a 1932 speech by R.S. Grant who went on later to captain West Indies: *Sporting Chronicle Souvenir Annual* (Port of Spain, 1932), p. 41.
44. Section based on *Agricultural Reporter*, 11 Sept. 1891.
45. E.C. Jackman, Letter to Editor, *Advocate*, 20 Jan. 1909, 3.
46. BCA 1904–05, 34–6; *Advocate*, 16 March 1905, 7.
47. For the English connections between sport and education through the period, J.A. Mangan, *Athleticism in the Victorian and Edwardian Public School; the Emergence and Consolidation of an Educational Ideology* (Cambridge, 1981). For the Barbadian 'education through cricket' philosophy, E.D. Laborde, 'Public School Cricket', *BCA* 1910–11, 108–13.
48. *Barbados Blue Book*, op. cit.
49. Lodge vicissitudes may be followed in the Governing Council Minutes, 1882–1914, BA.
50. A convenient reference for the background and growth of Harrison is *The Harrisonian: 250th Anniversary Commemorative Issue* (Bridgetown, 1983).
51. Somers Cocks is recalled in F.A. Hoyos, *Our Common Heritage* (Bridgetown, 1951).
52. For example, 'A very brilliant and treasured performance' saw Harrison's last-wicket partnership beat Spartan in 1895; the Lodge win over Wanderers in 1897 'caused a

sensation in town for days', and the Harrison victory over Wanderers the same season was 'hailed with delight by the crowd': *BCA* 1894–5, 85 and *BCA* 1897–8, 148 and 108.

53. The theory of cultural reproduction comes from Pierre Bourdieu. See his 'Cultural Reproduction and Social Reproduction' in Richard Brown (ed.), *Knowledge, Education and Cultural Change: Papers in the Sociology of Education* (London, 1973); with Jean-Claude Passeron, *Reproduction in Education, Society and Culture* (London, 1977), 'Sport and Social Class', *Social Science Information*, 17 (1978). For application to the educational area, Roger Dale *et al*, *Education and the State, II: Politics, Patriarchy and Practice* (London, 1981).
54. Brian Stoddart, *Saturday Afternoon Fever: Sport in the Australian Culture* (Sydney, 1986).
55. Material drawn from *The British Guiana Directory and Almanack*, op. cit.
56. J. Coleman Beecher, *Jamaica Cricket, 1863–1926* (Kingston, 1926); Herbert G. Macdonald, *History of the Kingston Cricket Club* (Kingston, 1938).
57. For a general view of the development of social divisions within Barbadian cricket, L. O'Brien Thompson, 'How Cricket Is West Indian Cricket?: Class, Racial and Colour Conflict', *Caribbean Review*, 12 (1983).
58. *BCA* 1899–1900, 110.
59. The Hinkson–Deighton wedding was reported as a great society occasion and drew the largest crowd ever to that time at St Michael's Cathedral, *Globe*, 13 Nov. 1893.
60. A description of Bancroft is in *Monthly Illustrator*, XV (1897).
61. For the spirit, *BCA* 1902–03, 46 and 1903–04, 61; for the incident, *Agricultural Reporter*, 23 Aug. 1895.
62. Personal information conveyed to the author.
63. Governor to Secretary of State, 3 Feb. 1900, No. 32, COL 2/1/44, BA.
64. For mention of the crowds, 'clans' and excitement: *Agricultural Reporter*, 13 Dec. 1895; *BCA* 1898–99, 65–68; *Bulletin*, 4 Dec. 1899, 8–10; *Advocate*, 5 Jan. 1909, 2.
65. *BCA* 1895–96, 132.
66. *Globe*, 20 Aug. 1891.
67. For an account of him playing for a ground staff team in 1893: *Globe*, 28 Aug. 1893.
68. Account based on *BCA* 1899–1900, 31–33 , and 129–132; *Bulletin*, all in 1899: 17 June, 8, 31 July, 8, 2 Sept., 16, 16 Sept., 12, 2 Oct. 11; *Globe* 1899: 19 July, 31 July, 8 Sept.
69. Pickwick and Windward reactions seen conveniently in *Bulletin* 1899: 15 Aug., 8 and 9, 19 Aug., 8, 24 Aug., 9–10, 25 Aug., 8–9, 31 Aug., 8–10, 1 July, 10–11, 4 Dec., 10, 9 Dec., 14.
70. *Globe*, 4 Aug. 1899.
71. *Agricultural Reporter*, 3 April 1900, 3.
72. *BCA* 1912–13, 141–142.
73. *Queen's Park Cricket Club Diamond Jubilee*, op. cit.
74. For ground strengths, *BCA* 1895–6, 189, and 1901–02, 158. The 1893 match, *Globe*, 28 Aug. 1893.
75. *BCA* 1903–04, 157; *BCA* 1913–14, 104.
76. Genealogical material drawn from Births, Deaths and Marriages, Parish Registers, BA.
77. *BCA*: 1899–1900, 170–171; 1900–1901, 72–4; 1904–05, 29; 1909–10, 27–34.
78. The best expression of this view is by 'Incognito' in *BCA* 1899–1900, 129–132.
79. *Bulletin*, 18 Aug. 1899, 8.
80. *BCA* 1902–03, 186. For a idealized view of the Frame Food heritage, Clyde A. Walcott, 'The Home of the Heroes', *New World Quarterly*, III (1966–67).
81. *Advocate*, 4 Jan. 1908, 5. There is the further point of an 1899 report claiming 600 members of island cricket clubs, *Globe*, 12 July 1899.
82. *Bulletin*, 11 Oct. 1899, 8.
83. The membership of the original Committee appears in *BCA* 1902–03, 186.
84. The report in *BCA* 1904–05, 158 saw 'cause for fear that bats and wickets would be utilised for other purposes than negotiating the ball'.
85. *Agricultural Reporter*, 11 Sept. 1891.

86. *Herald*, 25 April 1895.
87. *Advocate*, 30 Jan. 1905, 5.
88. *Agricultural Reporter*, 22 Jan. 1895.
89. Ibid., 7 Feb. 1899, 3.
90. Pelham Warner, op. cit., pp. 42–6; *Bulletin*, 3 Dec. 1897, 4–5.
91. See the Hesketh-Prichard report, *Advocate*, 22 March 1905, 9–11.
92. Andrew Ernest Stoddart, England cricket and rugby union captain, 1897 tourist; *BCA* 1896–97, 187.
93. *Agricultural Reporter*, 19 Jan. 1900. For comparative Caribbean expressions of popular content with local victories: Sir Reginald St Johnston, *From a Colonial Governor's Notebook* (London, 1936), p. 152, and Lord Hawke, op. cit., p. 169.
94. Bowen, op. cit., p. 16.
95. *Advocate*, 22 March 1905, 9–11.
96. *BCA* 1900–1901, 81–82.
97. *Globe*, 29 Sept. 1899.
98. Such as the 'much merriment' expressed when bowler Clifford Goodman struck English batsman F.W. Bush in 1897, *Bulletin*, 14 Jan. 1897, 9.
99. *BCA* 1908–09, 46–47.
100. Sir John Hutson, *Memories of a Long Life* (Bridgetown, 1948), p. 23.
101. *Bulletin*, 29 Sept. 1897, 9; *Globe*, 29 Sept. 1899.
102. *Bulletin*, 14 Jan. 1897, 8. My thoughts on colonial resistance have benefited from discussions with Ranajit Guha and from his book, *Elementary Aspects of Peasant Insurgency in Colonial India* (Delhi, 1984).
103. Bowen, op. cit., pp. 2 and 14; *Agricultural Reporter*, 29 Jan. 1895.
104. For these and other disturbances that year: *West Indian Royal Commission*, op. cit., p. 18; *Herald*, 1895: 4 April, 11 April, 12 April; *Agricultural Reporter*, 1895: 9 July and 16 July.
105. *Bulletin*, 26 Aug. 1899, 8; *BCA* 1903–04, 157.
106. The Griffith story is told briefly in John Wickham, 'Herman', *West Indies Cricket Annual* (Bridgetown, 1980), pp. 12–14.

14

Cricket and Colonialism: Colonial Hegemony and Indigenous Subversion?

Richard Cashman

The subject of cricket and colonialism has attracted much recent attention and there is a growing body of literature on the importation and adaptation of this game to various, mainly colonial, societies. The central focus has been why the game flourished in some societies but was largely rejected in others.

The studies include Mandle's seminal article on cricket and Australian nationalism, which explores the reasons for cricket's popularity there, and his study of the Gaelic Athletic Association, which explains why the Irish rejected English games, including cricket. The theme of rejection is taken up in Tyrrell's study of the rise of North American baseball: he looks at why cricket first took root but was later supplanted by baseball. The West Indian enthusiasm for cricket is well documented in C.L.R. James' much-acclaimed study, *Beyond a Boundary*. More recently Tiffin has looked at some of the implications of this perspective, such as whether the radical nationalist James had a blind spot in accepting the colonial game of cricket. A central concern of my own study is as to why cricket has assumed the proportions of a craze in India. A recent publication by Sissons and Stoddart on the 'bodyline' tour includes a chapter on 'The Empire of Cricket' which explores the emergence of the colonial ideology of cricket in a number of societies. Another book with some excellent exploration and interesting comments on cricket and imperialism is Birley's *The Willow Wand* on some cricket myths. Finally a great deal that is pertinent is now being written on sport and colonialism in general, such as Mangan's work on games and imperialism.[1]

In addition, there are numerous assertions and myths about the imperial role of cricket in the very extensive literature of the game. Ranji, for instance, claimed in the 1890s that cricket 'is certainly amongst the most powerful links which keep our Empire together'; Lord Harris went even further in 1921 when he claimed that 'cricket had done more to consolidate the Empire than any other influence'.[2] The distinguished historian G.M. Trevelyan could not resist the urge to elevate cricket when he contended that 'if the French noblesse had been capable of playing cricket with their peasants, their châteaux would never have been burnt'.[3] Numerous other writers have made the rather absurd statement that if the Germans (or Italians, or whoever) had played cricket there would have been no world war.

In the recent literature on the acceptance/rejection/adaptation of cricket in colonial societies, historians have focused on five main categories of explanation. The first, the ideology of nationalism, is one of the more prominent. Mandle has stressed that while nationalism was an important reason why Australians took up cricket so enthusiastically, it was also a central reason why the Irish rejected cricket. Mandle and others have contrasted the more thoroughgoing anti-colonial nationalism in Ireland with the love–hate type of nationalism evident in Australia, India and the West Indies. Tiffin has developed this idea further when she argued that James did not subject the colonial institution of cricket to the same radical scrutiny as other political institutions, presumably being too indoctrinated by the effectiveness of British promotion of cricket.

Secondly, Tyrrell and others have stressed the pivotal role played by élite groups and the significance of their relationship with client groups. Cricket in India was taken up, for various reasons, by socially important groups such as the Parsis of Bombay, and the spread of the game can be explained partly by the emulation of their behaviour by groups below. Tyrrell draws on the same idea to explain the demise of cricket in North America. Cricket took root there among some of the Anglophile-inclined élites of Philadelphia. Groups below them, such as German, Irish and Italian Americans, did not attempt to emulate the behaviour of the Philadelphia élite.

Thirdly, Tyrrell points out some of the dangers in adopting a cultural explanation for the rise of baseball at the expense of cricket. He is critical of those American historians who believe that baseball became popular because Americans wanted a faster and less time-consuming game more in tune with some emerging national culture. Other historians, however, do make use of cultural explanation (cricket in India fitted in with Hindu notions of pollution and purity whereas rugby did not), though they avoid some of Tyrrell's criticism by linking culture with particular groups.

A fourth explanation which fitted the Indian, and possibly the West Indian, situation, was the dearth of leisure alternatives, or the absence of a traditional game which could be transformed into a mass spectator sport.

The indigenous appropriation (domestication) of cricket represents a final area of explanation. Historians have unearthed many examples of this theme, ranging from the barracking tradition in Australia, to garlanding of centurions and the emergence of a curious hybrid commentary language of 'Hinglish' (English cricket terms located within the Hindi-language broadcast). A number of historians have placed considerable weight on this argument to account for the growth in popularity of the game in various societies.

Behind the numerous minor differences and shades of emphasis from study to study lie two major differences of interpretation and approach. There are those, first of all, who stress the power of cricket and its associated ideology to indoctrinate colonial subjects. Because it was wrapped up in the garb of a compelling and fascinating game, it was a very powerful imperial weapon. This view can be traced right through from Ranjitsinhji and Lord Harris to modern exponents, such as Tiffin, who argued that James' non-rejection of cricket represented a decided weakness in his anti-colonial stance.

Mangan, as much as anyone, has put forward a cogent case for the proselytizers in articles such as 'Eton in India'. In this article he demonstrates that imperial diffusion of the games ethic was accomplished with the establishment of colleges such as Aitchison, Daly, Mayo and Rajkot. He is undoubtedly right to stress 'the eventual success of Macnaghten and his proselytizing successors in smoothly translating an education ethic from England to India' and to argue that the educational ethic 'in only gently modified form' survived the imperialist.[4] The games ethic fired the imagination of a succession of princes – Ranji, Duleep, the Pataudis, the Patialas, Vizianagram and many others – who played very significant roles in the spread of British games in India along with many of the cultural values associated with the game.

In his more recent work on the spread of the games ethic to Africa ('Tom Brown in Tropical Africa'),[5] Mangan has enhanced the approach to the study of imperialism and games by exploring some aspects both of colonial proselytization and the often paradoxical and contradictory black African responses. He demonstrates that while black pupils were keen to respond to academicism promoted in the public schools they responded with far less enthusiasm to athleticism. This is an interesting argument and raises some pertinent questions, yet to be tackled, about why games proselytization works well in some societies, and at some times better than others, and why some aspects of colonial ideology are more accepted than others.

However, for those who look at cricket and colonialism more 'from below', or from the perspective of the recipients of the games ideology, there are significant problems with the perspective which concentrates exclusively on the proselytizer. Where does the promoting hand of the colonial master stop and where does the adapting and assimilating indigenous tradition start? Is it merely adaptation and domestication or does it go beyond that to constitute resistance and even subversion? And how far can the colonial acceptance of cricket be seen as superior colonial salesmanship or a successful exercise of social control using the highly developed and subtle ideology of games and colonialism? Or was it that many colonial subjects chose to pursue a game, because of the ideology, or even in spite of it, because it suited them to take up cricket for their own reasons? Or was the ideology of colonialism the starting point for the adoption of cricket but once the game was launched other factors came to bear which led to its spread and consolidation?

The film, *Trobriand Cricket*, provides proof, if proof were necessary, that the view 'from above' and colonial salesmanship do not tell us the full story. Cricket in the Trobriand Islands was introduced from above, by missionaries. The film shows in a quite remarkable way how this British colonial team game, which emphasized competition, discipline, and various public-school and imperial values, was totally transformed to reflect tribal values: so that everyone could play on each side (each team had about 40 players), not just the best eleven, the home side was the one which should always win (but by a modest margin) and there were appropriate tribal dances at the fall of each wicket. The film shows us the end product of indigenous adaptation, and does not illuminate the process by which this occurred.[6] So, quite clearly, while games are an effective vehicle for proselytization in some circumstances, they can be subverted in others.

This leads us to the second major problem. Some of those who have concentrated on the proselytizers have tended to view colonial salesmanship as a monolithic activity, and have minimized conflicts, generated within colonial society, by the imported games. Mandle's pioneering work on 'Cricket and Australian Nationalism' illustrates this point.[7] Not surprisingly he focused on the most observable arena of conflict, that of the all-powerful motherland providing a convenient symbol of attack for the unsure adolescent society. There are several problems with such an approach. As a cultural nationalist approach to history it assumes that there was a culturally monolithic response in sport to the imperial presence and ideology, that colonial society was united in its acceptance of a sporting nationalism. This approach also seems to minimize possible conflict within the colonial society itself, between local individuals (such as officials) who were totally in sympathy with the imported colonial ideology of games and those such as participants (who were often divided in their commitment to the imported games ideology), commentators, and spectators and business interests.

Mandle would have confronted this problem had he extended his argument to include a fifth stage of cricket and Australian nationalism in the period 1901–12. Not long after Federation, in 1905, the Board of Control was established. It was a more powerful governing association than any of the previous local, state and national associations and it provided officials with more potential control over the game. Not surprisingly a major rift soon emerged between officials and players, which came to a head in 1912, over power and profits. However, behind this conflict was a fundamental difference of opinion between the more middle-class amateur officials and the players who were partly motivated by more professional motives. So it seems clear that while a sport could provide a focus and symbol for an emerging idea of nationhood (and Mandle is rightly acknowledged for his seminal article on this point), the very process of nationalism and the setting up of national bodies also provided arenas of conflict. Mandle's perspective can only be enhanced by adding aspects of conflict in addition to consensus.

Some recent studies have benefited from this added dimension, Sandiford has explained cricket's eclipse by soccer as a mass spectator sport in England by the class attitudes of many cricket officials who doggedly refused 'to sell the game to the rising urban proletariat as aggressively and as effectively as the first division soccer clubs' and sometimes raised prices deliberately to restrict the size of the working-class crowd.[8]

There are many possible areas of conflict which await exploration: the relationships between the many constituent elements involved in a sport including colonial ideologists and officials and their representatives in colonial society, players, amateur and professional, journalists, business interests, spectators, and society at large. It is at this point important to reiterate that it is not intended to argue that one approach to cricket and colonialism is superior to another; rather it is to suggest that neither view tells us the full story about the introduction and adaptation of games in colonial societies. It is quite obvious that proselytization occurred and was successful *to some extent* but that the same could be argued of colonial 'resistance' or 'domestication'.

The problem has a number of dimensions. There is, first of all, the question the success and failure rate of proselytization – its balance sheet. There is, in addition, the 'grey' area between proselytization and resistance/domestication where the values behind a game may be subtly resisted and changed almost imperceptibly; what might be called the infiltration and subtle subversion of the games ethic. None of these arduous questions will be tackled here. Rather, an attempt will be made to focus on the more modest and practical question as to how one can define an indigenous adaptive and assimilative tradition, and whether this amounts to adaptation simply in the sense of domestication, or whether it amounts to resistance or even subversion.

II

It must be admitted quite frankly that it is far more difficult to attempt to look at the objects of proselytization rather than the proselytizers themselves. Those who, like Mangan, in his earlier works, look at the spread of the games ethic, whether in the English or colonial public schools, can focus on particular individuals and explore their stated ideology. The schools themselves provide convenient arenas in which the social historian can develop hypotheses about the degree to which social control operated.

Those working 'from below' have no individuals, institutions or clear and convenient ideological statements of alternative views on the games ideology or of reaction to proselytization. Worse still, subversion has mostly to be inferred from behaviour which is often reported through eyes more attuned to the proselytizers. This is a familiar enough problem for anyone working in the area of popular culture. However, while the problem is a tricky one, it is none the less one which has to be confronted because it also faces the historian writing 'from above'. Mangan, for instance, has demonstrated that within the framework of the imperial college a significant number of the pupils were impressed by the games ethic, and some of its leading graduates, notably Ranji, accepted enthusiastically both the games and the associated ideology. Ranji was, in other words, well proselytized or maybe amenable to proselytization. Mangan, however, would be the first to admit that four imperial colleges represent only a small part of the front line of the diffusion of the games ethic to Indian society generally. So the question must be posed, even for 'historians from above' as to what happened to the games ethic outside these colleges. Or, what occurred when the 'converted' attempted to apply the games ethic outside the narrow arena of a school or a similar institution where the ideology of games could be inculcated effectively.

Even a cursory examination reveals that games and the games ethic were significantly adapted when they were taken out of the closed environment of the school to wider society with its different values and demands. Ranji, in some ways, was an exception among the princely 'converts' to cricket in that he always remained a 'pure convert'. He absorbed the ideology of games so well that he became one of the leading apologists of the imperial benefits of cricket as well as being a symbol of amateurism in cricket's 'golden age'. However, many other princely products of 'Eton in India' drew rather more on the cricket traditions of eighteenth-century gentry than of those refined in the English public schools when they promoted games in their own states. Cricket, to them, was a convenient way of conspicuous consumption, of enhancing princely politics, of extravagant and ostentatious leisure and general intrigue. There are numerous and colourful stories about princes being carried to the wicket, bribing umpires and players, and using the game for personal aggrandisement. Outside the school the values associated with the games ethic were adapted to suit the whims of individual princely states. So while many of the princes were well proselytized, in that they developed a great passion for cricket, it is also true that they used cricket for self- and state-aggrandisement rather than for any commitment to the colonial games ethic. Clearly there is a need to explore some of the wider patterns of the diffusion and reception of the games ethic. It is an important area of study, both for those who want to understand the imperial spread of the games ethic, and for those who are more interested in the perspective of the colonial subjects, the proselytized.

In my work on Australian cricket crowds I have attempted to explore, among other things, some aspects of an alternative tradition or a local response to an imported ideology by focusing on the crowd. To explore this tradition fully one would want to look well beyond the crowd to consider schools, the media, clubs, business interests, officials and so on. None the less the crowd is a convenient starting point in that it includes a wide cross-section of those interested in the game, and in some ways is a microcosm of the local cricket world itself. Crowd behaviour, it could be argued, does reflect something of wider social attitudes to a game.

It is worth considering two traditions of Australian crowd behaviour which were prominent in the nineteenth century, gambling and barracking, because both were widely criticized as being antithetical to the morality of cricket, or the middle-class ideology associated with the games ethic. They are also of interest because while the history of gambling was a very straightforward case of successful suppression in the 1880s, barracking, and associated violence, was not amenable to official suppression and continued into the twentieth century as one of the significant 'underground' and, in some people's minds, 'subversive' traditions associated with the game. Just how 'subversive' barracking was is a central concern of this chapter.

Gambling was an established feature of Australian crowd behaviour until the 1880s, which meant that this English eighteenth-century tradition lingered on longer in Australia than in England where efforts to clean up the game occurred much earlier. During the 1850s Australian newspapers regularly advertised the odds and until the late 1870s bookmakers were housed in the grandstands at major matches.

During the 1870s gambling came into disfavour because of its associations with two riots. A close game in an 1873 Melbourne Final, witnessed by W.G. Grace, resulted in an invasion of the ground which was in part caused by betting. The most celebrated gambling-linked riot, however, occurred in 1879, when the English captain Lord Harris was assaulted by an individual member of a mob, estimated at 2,000, which swarmed on to the ground following an unpopular umpiring decision. Reports in the various papers suggested that the larrikins who led the charge, were urged on by 'a well-known and ill-favoured bookmaker' who had backed the local side to the extent of nearly £1,000.

Gambling, which was thought to be a prime reason for the violent crowd behaviour, was roundly criticized not only by the English captain, Lord Harris, an apostle of the amateur ethic who believed that gambling was a great blot on the purity of cricket, but also by the local middle-class administrators and the press. There were few defenders of the gamblers, the bookmakers and of those in the crowd who enjoyed gambling.

Gambling was virtually removed from Australian cricket in the 1880s because there was a solid consensus among those who ran the game, in two countries, that it was against the spirit of the games ethic. It was not difficult to achieve this result because Australians had many other gambling outlets and gambling could be curtailed simply by removing bookmakers from the pavilion and even the ground itself.

Barracking is a far more complex story. Australian barracking is probably as old a tradition as cricket itself. Barracking, in Australia, has always taken several forms. One practice was to jeer at or taunt the opposition, either in a vindictive or humorous manner, but it has also been used in the sense of support for the home side. Although Australians have their particular barracking idioms and famous barrackers, it is hardly a unique custom in that working-class crowds at football and cricket, and American crowds at baseball have all developed similar traditions. In many respects barracking can be seen as more normal crowd behaviour rather than the studied silences which were more highly regarded at cricket matches in the south of England. The only really unique aspect of Australian barracking is its idiom, the distinctive language and humour involved.

Until the 1880s officials in Australia were not greatly concerned with the practice of barracking. However, from about that time reports start to appear criticizing the Australian barracker, or at least, some individual barrackers. In 1883 a Melbourne paper reported that an unpleasant feature of a Test match was 'a cowardly attempt by a small knot of roughs to prevent Englishman Barnes catching the Australian Garrett, by jeering at the fieldsman as he was preparing for the catch'. The author of this article took consolation from the fact that this 'unmanly behaviour' of 'these small-souled fellows was far outweighed by the sporting behaviour of the majority of patrons'. The incident is a revealing one.[9] This Melbourne journalist argued that a small section of the crowd was not behaving according to the imported games ethic, in that they were attempting to intimidate the opposition. There is a great irony also in this comment for several Australian spectators were criticized for a form of behaviour pioneered by no one less than Dr W.G. Grace himself, a master sledger. Such behaviour today would hardly raise an eyebrow. More than a decade later the influential commentator J.C. Davis, writing in the *Australian Cricket Annual 1895–96*, deplored the increase in personal barracking over the previous few years which he described as 'regrettable, reprehensible and selfish' and was prompted by partisan feelings and often by 'a sheer desire to be abusive'.[10]

Barracking became a major issue of debate in the 1890s because it touched on a wide variety of issues relating to spectator behaviour. While everyone enjoyed the humorous comments of self-made barracking wits, others questioned the right of a paying spectator to attempt to 'intimidate' the opposition. Officials, too, soon came to realize that barracking could sometimes lead to a crowd demonstration, and that hooting could lead to throwing rubbish on to a pitch, and ultimately to an invasion of the pitch and the disruption of the game. Barracking then ran the gamut from seemingly harmless intrusion in the game to actual disruption.

Barracking up to this point was largely a local debate between those who ran the game and those who wrote about the game, the largely middle-class amateur-minded officials, and the paying spectators in the outer sections who had different notions of the purpose of play. However, during the 1897–98 tour it was elevated to an international problem in that the defeated England captain, A.E. Stoddart, came to view barracking as a type of colonial disease, presumably stemming from convictism, which was totally alien to the true ethic of games. The possible influence of convictism on Australian player behaviour (and presumably by inference crowd behaviour too) was made by the former England captain Ted Dexter as recently as 1972: writing in *Wisden* Dexter stated that: 'I have on occasions taken a quite unreasonable dislike to Australians . . . Under provocation . . . Australians [the players] . . . throw off all their 180 years of civilized nationhood; they gaily revive every prejudice they ever knew, whether to do with accent, class consciousness or even the original "convict" complex and sally forth into battle'.[11] This debate has emerged quite regularly right up to the present and variations of this theme have been taken up by numerous cricket luminaries such as Douglas Jardine in the 1930s, John Woodcock in the 1980s and many others.

During the 1897–98 tour the barracking question blew up for the first, and certainly not the last, time. By the end of the series the England captain Stoddart could not hide his contempt for the 'evil barrackers' who were 'no good to man or beast'. Stoddart argued that the crowd had deliberately set out to undermine his players by sustained comments and jeering and had succeeded on some critical occasions in 'getting under the skin' of some of his players, notably fast bowler Richardson 'who lost his head and his bowling'. Stoddart contended that the crowd played a significant role in the English defeat. That point today hardly worries many followers of sport, who recognize that home ground advantage is worth so many points or runs or goals, but in the era when the public school ideology was predominant, at least in middle-class games, crowd partisanship, hostility and even violence against the opposition was regarded as unfair and unsportsmanlike.[12]

One interesting aspect of the 1897–98 barracking debate and subsequent ones up to 1932–33 was that local officials tended to agree with this English criticism of Australian crowds. Some pressmen and even a few players, but not all, also echoed similar views. Australian officials, also subscribers to the amateur ethic, were embarrassed by the behaviour of their crowds in the early decades of the twentieth century and joined the general chorus against barracking, jeering, disturbances and invasions which were reasonably common until 1914.

The Australian press and players were not so much affected by the great national pastime of the 'cultural cringe'. In a very well-argued article in 1912 former Victorian and Oxford University player, P.R.

Le Couteur, contended that part of the English hostility to barracking arose out of a misunderstanding of Australian working-class behaviour and particularly their sardonic and self-deprecatory humour. He pointed out that by allowing comments to hurt them, instead of laughing at them, the England players made the situation far worse.[13] In the first decade other pressmen and players came out in defence of the Australian outer crowd, pointing out that they followed the game very intently and far more so than the ladies and gentlemen of the stands. Some of their comments arose out of disappointment, rather than disparagement, with the performance of a particular player. The Australian outer crowd, above all, come to the ground ready 'to play', to be involved in the drama itself. It was true that they were over-exuberant on occasions and some individuals overstepped the mark but on the whole they were 'well behaved'.

Cricket's barracking debate surfaced as a major issue in 1932–33 when the famous bodyline tour brought to a head a long-running debate on games morality. During this debate local officials, for the first time, attempted to develop a separate line of argument from their MCC 'overlords' by contending that the 'sportsmanlike' behaviour, the core of the games ethic, was being subverted by Jardine and his team when they introduced the tactic of 'bodyline' bowling: winning had become far more important than the morality of play. The games ethic was being subverted by the proselytizers themselves. While stonewalling on this particular line, by not admitting to subversion, the MCC, and Jardine in particular, developed a counter-attack. It was the claim that Australian officials, press and players tolerated a form of unsportsmanlike behaviour among the crowd (barracking) which was itself subversive of the spirit of cricket.

This was undoubtedly a great furore. Most objective commentators have concluded that the Australian crowd behaved with considerable discipline given the tension associated with this series. It also must be remembered that Jardine never tried to disguise his distaste for the crowd and thus antagonized them. It is surprising that his attitude did not generate more hostility.

There is no point here relating yet again the many nuances of cricket's most famous debate other than to point out that it had substantial implications for past and future debates about the ideology of games. It created a major split among the hegemonists, officials in England and Australia who had usually agreed on questions related to the morality of play. There was also a division of opinion among the English players with a minority sympathetic to the Australian side of the argument. There were always some Englishmen who took the

Australian 'side' or did not completely toe the English line in barracking debates.

Cricket's 1932–33 crisis was resolved when the Australian Board of Control backed down in that it was unwilling to pursue its argument that the English play was 'unsportsmanlike'. After the crisis relations between the two national authorities were patched up and their joint 'hegemony' was restored. This and earlier crises over barracking provide good case studies of what Raymond Williams would see as the complex and often fluid nature of alliances associated with the renewal, recreation and defence of games hegemony.

Criticism of Australian barracking and crowd behaviour has continued to surface quite regularly from time to time. More recently New Zealander Glenn Turner likened a tour to Australia as being akin to a posting to [war-time] Vietnam. No less an authority than *Wisden* editor, John Woodcock, after witnessing a one-day night cricket game on 12 January 1982, was scathing in his attack on the Sydney crowd. 'England,' he wrote, 'was drowned in a sea of jingoism in yesterday's World Series Cup . . . By 8.30 any resemblance between what was happening and any normal game of cricket was coincidental . . . The sounds of fury, the beating of boards, and the booing of Englishmen were orgiastic. This was not so much sport as jingoism'.[14] The crowd, in Woodcock's opinion, were not only 'discordant and unattractive' but were subverting the game.

Although barracking often amounted to nothing more than humorous diversion, it none the less was an example of a continuing alternative tradition of behaviour. For the most part it did not represent a particularly subversive tradition in that it did not challenge the power or hegemony of the ruling oligarchy. It was more of a nuisance in that the violence associated with barracking was often difficult to control. However, it did represent an alternative form of behaviour which had to be tolerated in that it was impossible to stamp it out. Barrackers were people who openly expressed their ideas about the game, which frequently did not fit in with the games ethic.

Officials, the press, players, and businessmen – all those associated with the promotion of cricket – had to deal with barracking in one way or another and to work out their views on this alternative tradition. In defending the Australian crowd in the 1930s, players, press and officials were redefining acceptable behaviour at cricket, and in a subtle way, redefining the ideology of the game in an Australian context.

It is a very long jump from 1932–33 to the World Series 'takeover bid' of 1977 when a media magnate, Kerry Packer, persuaded the bulk of the best players of the world to join him in an alternative cricket league. World Series Cricket (WSC) introduced not only new structures and forms of competition but was also based on a new moral emphasis relating to play: greater professionalism, commercialism and commitment to entertainment were central ideas. The success of WSC, and the subsequent elevation of the one-day contest, was based, at least in part, on a very shrewd assessment of an Australian audience preference for a more volatile, noisy and even violent game in which winning (rather than just drawing) was important. Partisanship and jingoism, rather than being seen as a vice, were encouraged in the interest of generating more crowd enthusiasm, and selling both more tickets for the game and products on television. Notions such as fair play and sportsmanship were discounted in favour of the new values of winning, 'aggro' and 'hype'.

It has become quite fashionable to see Kerry Packer as 'subverting the system from above' in that many have argued that 'he created a new audience for the game' and brought about a 'revival' in cricket through money, promotion and the media. This is partly a myth because Packer did not take over cricket at a time of weakness; rather there was a television-related boom in crowds from the 1970s.[15] Nor did Kerry Packer create the new mood of jingoism, or barracking, or the associated intimidation of opponents (and even the home players) all of which sometimes led to violence. The vast core of WSC patrons, including 'ockers' and 'yobbos' who became more prominent on the hills, were well to the fore and very vocal from the late 1960s on. In a sense this alternative attitude towards the game had taken root for well over a century. Much as I hate to argue it, being a cricket traditionalist with a preference for Test cricket and its associated games ideology, it seems clear that Kerry Packer plucked a more than century-old tradition and redefined it (and even manipulated it) to suit his own commercial interests.

It is popular today to blame Packer, Channel 9 (the television company of his association) and the players for what many see as the parlous state of the game in Australia: over-commercial, vulgar, 'unholy and profane' and 'trivial' (as Huizinga, Lasch, Tatz and company would see it). What is sometimes overlooked is that without the support of a significant element of the crowd WSC would not have got off the ground. While it is true that Kerry Packer manipulated the Australian tradition of crowd behaviour, and added many new television-directed elements, there is an element of continuity between an alternative tradition and the new cricket hegemony which has domesticated the crowd more than ever before.

III

This chapter does not explore all the subtleties of the relationship between those who are attempting to proselytize and those who are the subjects of proselytization. All that has been suggested is that relationships between component groups are complex and often change. It has also been argued that those who do not hold power in the administration of games can develop alternative values which not only can challenge the official morality of a game but can become part of a redefined hegemony, such as the WSC takeover. We need to know more about the changing relationship between the various component parts of the world which ran cricket before such assertions can be accepted.

This initial exploration of this subject suggests that the alternative tradition of barracking was not in itself very significant in power terms, the alternative ideology which it articulated was not seen as a threat by the authorities. So it was hardly a subversive tradition for the most part. However, the tradition did take on some importance when it was discovered by the counter-hegemonist, Kerry Packer, who elevated it to a tradition of great importance.

The task then, in my opinion, is partly to work out what went on outside 'Eton of India' and the like when the lessons of games were well taught and well learnt. Adaptation and change outside these arenas of games morality undoubtedly led to significant changes and subtle emanations in the creed. All of this is a very familiar theme to anyone who has studied the spread of religious or conversion movements which often begin as 'pure' movements but assume an altered form by the time they have fitted into the social structure and values of the new environment. Brotherhood is a central notion in the great tradition of Islam, but it has adapted to such an extent to the environment of India that there is a caste system within Indian Islam.

Future research will probably show that C.L.R. James did not have a blind spot at all when it came to cricket. It is likely to show that this game, which was exported to the West Indies as part of the ideological property of the colonial masters, was transformed to become a crucial element of emerging West Indian mass culture and identity. The task of analysing the colonial domestication of games, and of interpreting whether this represents adaptation, resistance and/or subversion, has only just begun.

272   BRITISH CULTURE AND SPORT AT HOME AND ABROAD, 1700–1914

NOTES

1. W.F. Mandle, 'Cricket and Australian Nationalism in the Nineteenth Century', *Journal of the Royal Australian Historical Society*, LIX, 4 (December 1973) 225–46; 'Sport as Politics: the Gaelic Athletic Association 1884–1916', in Cashman and McKernan (ed.), *Sport in History* (St Lucia, Qld, 1979), pp. 99–123; Ian Tyrrell, 'The Emergence of American Baseball c. 1850–80', in *Sport in History*, pp. 205–26; C.L.R. James, *Beyond a Boundary* (London, 1963); Helen Tiffin, 'Cricket, Literature and the Politics of Decolonisation – the Case of C.L.R. James', in Cashman and McKernan, *Sport: Money, Morality and the Media*, (Kensington, NSW, 1981), pp. 177–93; Richard Cashman, *Patrons, Players and the Crowd* (New Delhi, 1980); Ric Sissons and Brian Stoddart, *Cricket and Empire: The 1932–33 Bodyline Tour of Australia* (Sydney, 1984); Derek Birley, *The Willow Wand: Some Cricket Myths Explored* (London, 1979); see various articles by J.A. Mangan on the subject, such as 'Eton in India: The Imperial Diffusion of a Victorian Educational Ethic', *History of Education*, VII, 2 (1978), 105–18.
2. Birley, *The Willow Wand*, p. 13.
3. Ibid., p. 43.
4. Mangan, 'Eton in India', pp. 114, 118.
5. J.A. Mangan, 'Tom Brown in Tropical Africa', paper delivered at the International Scientific Conference of the Los Angeles Olympics, Oregon, July 1984.
6. Released in 1976 by the Papua New Guinea Office of Information. An Australian anthropologist helped to produce the film.
7. Mandle, 'Cricket and Australian Nationalism'.
8. Keith A.P. Sandiford, 'English Cricket Crowds during the Victorian Age', *Journal of Sport History*, IX (Winter 1982), 5–22; see also Wray Vamplew, 'Sports Crowd Disorder in Britain, 1870–1914; Causes and Control', *Journal of Sport History*, VII (Spring 1980), 5–50.
9. *Age*, 23 January 1883.
10. *Australian Cricket Annual*, 1895–96.
11. *Wisden* 1972, 91–96.
12. Richard Cashman, *'Ave a Go, Yer Mug! Australian Cricket Crowds from Larrikin to Ocker* (Sydney 1984), pp. 48–51.
13. *Referee*, 10 January 1912.
14. *Sydney Morning Herald*, 13 January 1983.
15. Richard Cashman, *Australian Cricket Crowds. The Attendance Cycle. Daily Figures, 1877–1984*, (Kensington, NSW, 1984), p. 17.

INDEX